T0135585

Bibliografische Information der Deutschen Nationalbibliothek

Die Deutsche Nationalbibliothek verzeichnet diese Publikation in der
Deutschen Nationalbibliografie; detaillierte bibliografische Daten sind
im Internet über http://dnb.d-nb.de abrufbar.

ISBN 978-3-8325-3870-5

Logos Verlag Berlin GmbH
Comeniushof, Gubener Str. 47,
10243 Berlin
Tel.: +49 (0)30 42 85 10 90
Fax: +49 (0)30 42 85 10 92
INTERNET: http://www.logos-verlag.de

Low-Power Galileo/GPS Single-Shot Receiver Architecture for Mobile Terminals

Low-Power-Galileo/GPS-Single-Shot-
Empfängerarchitektur
für mobile Endgeräte

Der Technischen Fakultät der
Universität Erlangen-Nürnberg
zur Erlangung des Grades
DOKTOR-INGENIEUR

vorgelegt von
Christoph Kandziora
aus Erfurt

Als Dissertation genehmigt von
der Technischen Fakultät der
Friedrich-Alexander-Universität Erlangen-Nürnberg
Tag der mündlichen Prüfung: 10.11.2014

Vorsitzende/r des Promotionsorgans: Prof. Dr.-Ing. habil. Marion Merklein

Gutachter/in: Prof. Dr.-Ing. Dr.-Ing. habil. Robert Weigel
                 Prof. Dr.-Ing. Stefan Lindenmeier

Die besten Dinge im Leben sind nicht die, die man für Geld bekommt.

Albert Einstein

Dedicated to Kim, Julius and Josephina

Acknowledgments

In the years of work on this thesis I have received support and encouragement from a great number of individuals. In particular I would like to thank Prof. Dr.-Ing. Dr.-Ing. habil. Robert Weigel for giving me the opportunity to create this work under his supervision. The freedom in scope while developing the ideas of this work as well as his technical input were extremely helpful. Additionally his patience and encouragement even after my employment at his institute, finally made it possible to finish this work.

I would like to thank Univ.-Prof. Dr. Mario Huemer who gave me the idea and raised my interest in working as a researcher at the institute.

For their technical input to this work I would like to thank Prof. Dr.-Ing. Georg Fischer and Dr.-Ing. Ulrich Tietze.

During my time at the institute many colleagues provided me support not only in technical discussions but also at great times having beer after and besides work, namely my former office colleagues Dietmar Kissinger, Jochen Rascher and Benjamin Waldmann as well as other colleagues like Jochen Essel, Alban Ferizi, Stefan Zorn, Benjamin Sewiolo, Errikos Lourandakis, Benjamin Lämmle, Markus Gardill and Gabor Vinci. Thank you guys for your support and the great times!

Additionally a big "Thank you!" to Gabriele Köhnen for the intense discussions after lunch break and your priceless support regarding administrative issues.

Last and most important, I have to thank my family for supporting me during the long stressful time while working on this thesis. I'd like to thank my parents and my wife Kim who often enough ensured that I was not troubled with other problems.

Last but not least I have to apologize to my wife and my children for my intellectual and personal absence during the work. And finally thank you, Kim, for getting up with the children!

Abstract

In the last few years satellite based navigation has become an integral part of everyday life. Starting in the beginnings with primarily military applications together with usage in the civil area only by academic and private enthusiasts, the triumph procession of navigation solutions for everyone accelerated in the early 2000s. With the deactivation of the selective availability of the Navigational Satellite Timing and Ranging-Global Positioning System (NAVSTAR GPS) system, reasonable accuracy and availability was guaranteed and allowed the development of reliable and useable navigation solutions. The drive of integration and miniaturization soon resulted in the appearance of navigation capabilities into mobile devices, especially cellular phones, making use of the advantages of cell phone communication in parallel to the positioning solution.

This doctoral thesis introduces an enhanced low-power architecture for an assisted Galileo/GPS receiver. The combination of time- and navigation signals eliminates the main drawback of assisted GNSS receiver architectures, its large baseband signal processing effort. Meanwhile, the advantages of well known assisted architectures, a short Time-to-First-Fix (TTFF) at low power consumption, are preserved. For the first time, the effort in digital baseband processing is compensated by the integration of a Long-Wave (LW)-time code receiver into a Global Navigation Satellite System (GNSS) receiver architecture. Although minor additional hardware in the analog domain for the reception of the LW signal increases power consumption this is by far compensated by the savings in power consumption due to the reduction in computational effort for the determination of the code phases of the Space Vehicles (SVs) in the digital baseband domain.

Based on a common assisted receiver architecture, the Single-Shot receiver architecture, the augmentation by the time information allows, together with the navigation message and an approximate position, the pre-calculation of the incoming satellite signals. The time code signal used in the present architecture is transmitted by the German DCF77 station.

While common GNSS RF-architectures can be reused, the reception of the DCF77 signal can be realized with minor additional hardware in the analog domain. Meanwhile in the digital domain, the engine originally implemented for the calculation of the actual GNSS code phases can be adapted for the determination of the DCF77 code phase. The excessive usage of processing power for the evaluation of the SV code phases can then be cut down to the amount required for DCF77 signal processing.

Only after time synchronization via the DCF77 signal, the pre-calculation of the incoming satellite signal is possible. A synchronization of better than 500 µs, which is guaranteed by the present architecture, not only eliminates possible ambiguities in the positioning solution but approves the assumption of having the greatest Pseudo-Random Noise Sequence (PRN) code phase search window for the time code signal. Therefore the pre-calcuation of the satellite signal results in a reduction of the search window for the actual code phase of the satellite signal by a factor of approximately 1.58 for the NAVSTAR GPS. Since the size of the resulting code phase search window does not scale with code length due to absolute time synchronization, the reduction in computational power when incorporating Galileo signals is even higher. Being four times longer than the length of NAVSTAR GPS codes the implementation of Galileo signals results in the reduction factor being increased to approximately 6.33. Since the determination of the satellite signal code phases is by far the dominant part of the baseband signal processing and therefore responsible for the dominant part of power consumption it is suitable to assume the same scaling factors in power consumption, too.

Although more complex architectures, for example two frequency approaches, can deliver higher accuracy, the targeted mobile platform for use in the mass market prohibit such extensions without denying the applicability of the concept. Instead a single-frequency approach is pursued here incorporating the GPS-L1-C/A signals for positioning.

The proof of concept is conducted by the implementation of the architecture in a system simulation, mainly performed in the Equivalent Complex Baseband (ECB).

Zusammenfassung

Die Entwicklungen des letzten Jahrzehnts haben der Satellitennavigation einen großen Aufschwung beschert. Dabei hat sich das vorwiegende Anwendungsgebiet von militärischen Applikationen und Nischenanwendungen für Enthusiasten zu einem Bestandteil des täglichen Lebens entwickelt. In den frühen Jahren des neuen Jahrtausends begann der Triumphzug von Satellitennavigationanwendungen für jedermann. Insbesondere durch die Deaktivierung der "Selective Availabilty"des NAVSTAR GPS- Systems wurden ausreichende Genauigkeit und Verfügbarkeit garantiert, die die Entwicklung von präzisen und damit sinnvoll nutzbaren Navigationsgeräten ermöglichte. Die fortwährende Integration und Miniaturisierung sorgten bald für das Auftauchen von Geräten, bei denen mobile Telefone mit einer Navigationslösung kombiniert wurden und von den Vorteilen der parallelen Kommunikation profitieren.

Diese Doktorarbeit stellt eine weiterentwickelte Architektur eines Assisted-Galileo-GPS-Empfängers vor. Die Kombination von Zeitsynchronisations- und Navigationssignalen eliminiert den größten Nachteil der Assisted-Architektur. Dabei wird der massive Einsatz von Rechenleistung in der Basisbandsignalverarbeitung reduziert. Trotzdem bleiben die Vorteile der bekannten Architekturen, eine kurze Time-to-First-Fix bei geringem Stromverbrauch erhalten. Zum ersten Mal wird dabei der Aufwand der Basisbandsignalverarbeitung durch die Integration eines LW-Zeitzeichenempfängers reduziert. Obwohl dabei zusätzliche Elektronik im Analogteil notwendig ist, die einen erhöhten Leistungsverbrauch mit sich bringt, wird dieser Leistungsverbrauch durch die Einsparung im digitalen Basisband mehr als kompensiert.

Aufbauend auf der bekannten Architektur des Single-Shot-Empfängers erlaubt die Erweiterung um die Zeitinformation zusammen mit der bereits bekannten Navigationsnachricht und der geschätzten Position eine Vorausberechnung der zu erwartenden Satellitensignale. Bei dem Zeitzeichensignal, das im Rahmen dieser Arbeit verwendet wird, handelt es sich um das Signal des deutschen DCF77-Senders. Während der für die analoge Vorverarbeitung der Navigationssignale zuständige Teil des Empfängers (GNSS RF-Frontend) unverändert übernommen wird, kann der Analogteil der Signalverarbeitung des DCF77-Signals mit geringem Aufwand realisiert werden. Währenddessen soll die digitale Signalverarbeitung, die für die Bestimmung der Kodephase der GNSS-Satellitensignale und des Zeitzeichensenders verantwortlich ist und dazu massiven Einsatz von Rechenkapazität erfordert, für die Korellation beider Signaltypen verwendet werden. Die Zeitsynchronisation durch das DCF77-Signal erlaubt die Vorausberechnung der ankommenden Satellitensignale. Eine Synchonisation der Uhrzeit, kleiner oder gleich 500 µs, wie sie durch die vorgestellte Architektur garantiert wird, eliminiert nicht nur mögliche Mehrdeutigkeiten in der Positionsbestimmung sondern zeigt ebenso, dass das größte Kodephasen-Suchfenster in der Berechnung der Zeitinformation zu finden ist. Die Vorausberechnung der Satellitensignale resultiert in einer Reduktion der Breite des Suchfensters für die augenblickliche Kodephase der Satellitensignale um den Faktor 1,58 bei Verwendung eines NAVSTAR GPS-Signals. Da die Größe des resultierenden Suchfensters aufgrund der absoluten Zeitbestimmung nicht mit der Länge des verwendeten Kodes skaliert, steigert sich der Einspareffekt der Rechenkapazität bei Verwendung von Galileo-Satellitensignalen. Auf Grund der viermal längeren Kodelänge im Vergleich zu GPS-Kodes wird der Einspareffekt auf den Faktor 6,33 gesteigert. Da die Bestimmung der Kodephasen bei Weitem den größten Anteil der Basisbandsignalverarbeitung ausmacht und damit auch für den größten Energieverbrauch verantwortlich ist, können die Einsparfaktoren für die Rechenkapazität nahezu unverändert für den Leistungsverbrauch angenommen werden.

Obwohl komplexere Architekturen, zum Beispiel Mehrfrequenzempfänger, höhere Genauigkeit liefern können, verbietet der Kostendruck im Bereich der Zielplatform eines mobilen Endgeräts für den Massenmarkt derartige Ansätze. Trotzdem wäre der hier vorgestellte Ansatz für eine Architektur diesen Typs ebenso geeignet. Stattdessen wird im Rahmen dieser Arbeit die alleinige Verwendung der GPS-L1-C/A-Signale verfolgt.

Der Nachweis der Funktionalität wird über eine Systemsimulation geführt. Diese wird aus Effizienzgründen im äquivalenten komplexen Basisband (engl. equivalent complex baseband, ECB) durchgeführt, ohne dabei die Gültigkeit der Ergebnisse einzuschränken.

Contents

List of Figures   v

List of Tables viii

List of Abbreviations   ix

1 Introduction   **1**

   1.1   Motivation   1

   1.2   Goals and Organization of this Work   2

2 Global Navigation Satellite Systems   **5**

   2.1   Principle of Satellite Navigation   5

      2.1.1   Positioning using GNSS Space Vehicles   5

      2.1.2   Reference Coordinate Systems   8

      2.1.3   Satellite Orbit and Propagation Error Correction Parameters . . .   9

   2.2   NAVSTAR GPS 13

      2.2.1   NAVSTAR GPS System Segments 14

      2.2.2   Signal-in-Space Generation and Characteristics 15

   2.3   Galileo 20

      2.3.1   Galileo System Segments 21

      2.3.2   Signal in Space Characteristics 21

3 State-of-the-Art GNSS Receiver Architectures   **27**

   3.1   Acquisition and Tracking Receiver Architecture 27

   3.2   Single-Shot Receiver Architecture 30

i

3.3   Search Space versus Time-to-First-Fix   34

4 Long-Range Time Synchronization   **35**
4.1   DCF77 Signal-in-Space Characteristics   36
4.2   Time Synchronization via LW Radio Signals   38

5 DAGNSS Receiver Architecture   **43**
5.1   System Overview   44
5.2   Analog-Radio-Frequency (RF) Frontend Architectures   45
       5.2.1   GNSS RF Frontend   45
       5.2.2   DCF77 RF Frontend   46
5.3   Digital Baseband Signal Processing Architecture   48
       5.3.1   Input Signal Memory   49
       5.3.2   Local PRN Memory   51
       5.3.3   Correlation Engine   54
       5.3.4   Baseband Control   58
5.4   Host Hardware and Operating System   58
5.5   Positioning Software   59

6 Systemsimulation   **71**
6.1   Simulation Setup   71
       6.1.1   DCF77 SIS Model   72
       6.1.2   GNSS SIS Model   75
       6.1.3   DCF77 RF Frontend Model   78
       6.1.4   GNSS RF Frontend Model   82
       6.1.5   Digital Baseband Signal Processing Model   83
       6.1.6   Positioning Solution   83
6.2   Simulation Results   84
       6.2.1   DCF77 Simulation Results   84
       6.2.2   GNSS Simulation Results   87

7 Summary and Outlook   **101**

List of Figures

2.1   2D-Positioning using two space based transmission beacons   6

2.2   3D-Positioning using three space based transmission beacons   7

2.3   Illustration of Keplerian elements defining the orientation of the orbit   . .   10

2.4   Constellation of NAVSTAR GPS SV [1]   14

2.5   Location of NAVSTAR GPS monitor stations   15

2.6   NAVSTAR GPS code generation   16

2.7   NAVSTAR GPS Signal-in-Space (SIS) generation   17

2.8   Frequency plan of existing GNSSs [2]   21

2.9   Format of navigation message frame   22

2.10 Constellation of Galileo [3]   23

2.11 Block diagram of Galileo signal generation   24

2.12 Spectra of Galileo E1 Open Service (OS) and NAVSTAR GPS Coarse Acquisition Code (C/A code) signals [4]   25

3.1   Block diagram of Acquisition and Tracking Receiver (ATR) architecture .   28

3.2   GNSS signal cross correlation properties   29

3.3   Block diagram of Single-Shot architecture   31

3.4   Normalized results of Cross Correlation Function (CCF) implemented in a Single-Shot-Receiver architecture (undisturbed scenario)   32

3.5   Block diagram of Frequency Domain Single-Shot Architecture   33

4.1   Behavior of DCF77 signal in amplitude and phase   36

4.2   Spectrum of the transmitted DCF77 signal (upper: spare antenna, lower: main antenna) [5]   38

4.3   Coverage of DCF77 signal (ground wave dominant and one ionospheric scattering)   39

4.4   DCF77 signal propagation   40

4.5   Pictures (a)(b) and equivalent circuit diagram (c) of DCF77 ferrite rod antennas [6]   41

4.6   Block diagram of common DCF77 receiver Frontend   42

5.1   Block diagram of newly developed Single-Shot-Receiver (SSR) based receiver architecture   44

5.2   Block diagram of DCF77 RF Frontend for Phase Shift Keying (PSK) demodulation   47

5.3   Dual-port organization of the input memory   50

5.4   First-In-First-Out (FIFO) based organization of input memory   51

5.5   FIFO based organization of PRN memory   52

5.6   Correlator channel implementing coherent and non-coherent integration .   54

5.7   Baseband signal processing correlator bank   55

5.8   Maximum finder circuit   57

5.9   Flowchart of software algorithm to determine the actual position   60

5.10 Code phase error approximation for satellites at high and low elevation angles   62

5.11 Different possibilities of false maximum definition of DCF77 CCF   64

5.12 Linear interpolator for CCF maximum: (a) Straight line through early and present points, (b) Intersection calculation, (c) Failure of exact code phase calculation, (d) Straight line through present and late points first   65

6.1   DCF77 TX antenna frequency response   73

6.2   Original and converted DCF77 RX antenna model   79

6.3   Results of the CCF for DCF77 signal simulation implementing either 1 bit(blue) and 2 bit(red) Analog to Digital Converters (ADCs)   80

6.4   Mixer output (red) compared to Filter output (green) (overlapping red graph) filtered with by lowpass characteristic (blue)   81

6.5   Mixer output (red) compared to Filter output (green) filtered with by lowpass characteristic (blue), zoomed to highlight baseband output   82

6.6   Worst-case assumption for inital position approximation error   85

6.7   Exemplary worst-case calculation for possible inital position approximation   86

6.8   Output of cross correlation function for worst case approximation of ini-
      tial position (solutions for points A (red), B(blue) and C(green) from Fig-
      ure 6.6   86

6.9   Correlation output, applying DCF77 RF-Frontend simulation with no dis-
      tortion effects   87

6.10 Correlation output, applying DCF77 RF-Frontend simulation with .1%
      jitter on Local Oscillator (LO) input of mixer, -25 dB SNR and interfer-
      ence signal with 3 dB higher signal power than DCF77 SIS   88

6.11 Output of CCF for SVs in view at boundary conditions given in table 6.3
      corrected by approximated phase shift of local PRN replica   90

6.12 Output of cross correlation bank (coherent integration time: 4 ms) for SV
      2, 10 dB SNR for DCF77 signal and $\frac{C}{N_0}$ = -21 db for GPS (red: output of
      correlation bank, blue: x100 sinc interpolation)   91

6.13 Focus on peak correlation output as given in Fig. 6.12 (red: output of
      correlation bank, blue: x100 sinc interpolation)   91

6.14 Preconditions and results (mean of 5 iterations) of positioning solution in
      an undisturbed scenario (corners of white line: individual results)   92

6.15 Preconditions and results (mean of 5 iterations) of positioning solution in
      an undisturbed scenario (corners of white line: individual results)   94

6.16 Preconditions and results (mean of 5 iterations) of positioning solution in
      an undisturbed scenario (corners of white line: individual results)   95

6.17 Results of the cross correlation function of the DCF77 signal at different
      noise levels   96

6.18 Maxima of the results of the cross correlation function of the DCF77 sig-
      nal at different noise levels   96

6.19 Results of the cross correlation function of the DCF77 signal at different
      noise levels   97

6.20 Results of the cross correlation function for SIS of the GPS SV2 at differ-
      ent noise levels (Bandwidth= 4 MHz)   98

6.21 Maximum of the cross correlation function for SIS of the GPS SV2 at
      different noise levels (Bandwidth= 4 MHz)   99

6.22 Calculated user position compared to real and approximate user position
      prior for different noise levels for GPS signals, including faulty reception
      conditions 100

List of Tables

2.1   Definition of Keplerian elements for NAVSTAR GPS satellite orbits . . .   10

2.2   Definition of ephemeris elements for NAVSTAR GPS satellite orbits . . .   11

2.3   Definition of ionospheric model parameters for the propagation of the
      L1-NAVSTAR GPS signals   12

2.4   Taps for SVs separation [7]   17

2.5   Subframe one parameters: specific satellite clock and health data [7] . . .   18

2.6   Subframe two and three parameters: specific satellite ephemeris parame-
      ters [7]   19

2.7   Subframe four pages: support data [7]   19

2.8   Subframe five pages: support data [7]   20

2.9   Subframe five pages: support data [7]   20

2.10 Galileo signals and corresponding carrier frequencies   23

2.11 Odd-numbered subframe of Galileo F/NAV message   26

2.12 Even-numbered subframe of Galileo F/NAV message   26

4.1   Time information encoded in DCF77 SIS [8]   37

5.1   Estimation of the number of CCF channels for determining exact code
      phase signals with different channel spacings and Doppler frequencies . .   48

6.1   Exemplary worst-case error assumption positions   85

6.2   Worst-case time errors compared to approximate inital position   87

6.3   Prerequisites of simulation in ideal reception environments   89

6.4   Simulated code phases for SVs in view assuming conditions according
      table 6.3   90

6.5   approximate versus calculated versus real user position   92

6.6   Prerequisites of simulation for Points A and B as given in Figure 6.6 . . .   93

6.7   Real versus simulated versus estimated code phases for SV 2 and 5   93

6.8   approximate versus calculated versus real user position for maximum
      faulty position approximation   94

6.9   Time synchronization and calculated code phases for DCF77 signal at
      several noise levels compared to calculated code phase for NAVSTAR GPS
      SV 2   97

6.10 Positioning solution and position error plotted versus noise level and real
      user position   99

List of Abbreviations

3GPP	3rd Generation Partnership Project
ACF	Auto Correlation Function
ADC	Analog to Digital Converter
AFB	Airforce Base
AGNSS	Assisted Global Navigation Satellite System
AM	Amplitude Modulation
ASIC	Application Specific Integrated Circuit
ATR	Acquisition and Tracking Receiver
AWGN	Additive White Gaussian Noise
BOC	Binary Offset Carrier
BPSK	Binary Phase Shift Keying
C/A code	Coarse Acquisition Code
CCF	Cross Correlation Function
CDDIS	The Crustal Dynamics Data Information System
CDMA	Code Division Multiple Access
CPU	Central Processing Unit
CRC	Cyclic Redundancy Check
CRT	Cathode Ray Tube

CS	Commercial Service
CW	Continuous Wave
DAGNSS	Double Assisted Global Navigation Satellite System
DDS	Direct Digital Synthesis
DST	Daylight-Saving Time
ECB	Equivalent Complex Baseband
ECEF	Earth-Centered-Earth-Fixed
ECI	Earth-Centered Inertial
EPL	Early-Present-Late
EU	European Union
FFT	Fast Fourier Transformation
FIFO	First-In-First-Out
FIR	Finite-Impulse-Response
FPGA	Field-Programmable Gate Array
FTP	File Transfer Protocol
GCS	Ground Control Segment
GMS	Ground Mission Segment
GNSS	Global Navigation Satellite System
NAVSTAR GPS	Navigational Satellite Timing and Ranging-Global Positioning System
GSM	Global System for Mobile Communications
GST	Galileo System Time
HID	Human Interface Device
HOW	Hand-Over-Word
IFFT	Inverse Fast Fourier Transformation
IF	Intermediate-Frequency
I&D	Integrate-and-Dump
IQ	In-Phase Quadrature
JFET	Junction Field Effect Transistor

LBS	Location-Based-Services
LFSR	Linear Feedback Shift Register
LMU	Location Measurement Unit
LOS	Line-of-Sight
LO	Local Oscillator
LUT	Look-up-Table
LW	Long-Wave
MAC	Multiply-Accumulate
MCS	Master Control Station
MW	Medium-Wave
NTP	Network-Time-Protocol
OS	Open Service
PM	Phase Modulation
PRN	Pseudo-Random Noise Sequence
PRS	Public Regulated Service
PSK	Phase Shift Keying
PTP	Precision-Time-Protocol
P/Y code	Y-encrypted Precise Code
QoS	Quality-of-Service
RAM	Random-Access-Memory
RF	Radio-Frequency
RINEX	Receiver Independent Exchange Format
ROM	Read-Only-Memory
RSSI	Received Signal Strength Indicator
RTC	Real-Time-Clock
SA	Selective Availability
SIS	Signal-in-Space
SNR	Signal-to-Noise Ratio

SoL	Safety-of-Life Service
SPI	Serial Peripheral Interface
SSR	Single-Shot-Receiver
SV	Space Vehicle
SW	Short-Wave
TDOA	Time Difference of Arrival
TLM	Telemetry Word
TOA	Time-of-Arrival
TOF	Time-of-Flight
TRANSIT	Navy Navigation Satellite System
TTFF	Time-to-First-Fix
USA	United States of America
UTC	Coordinated Universal Time
VEEPLVL	Very Early-Early-Present-Late-Very Late
WGS-84	World Geodetic System 1984
WLAN	Wireless Local Area Network
XOR gate	Exclusive-Or Gate

Für den gläubigen Menschen steht
Gott am Anfang, für den
Wissenschaftler am Ende aller seiner
Überlegungen.

CHAPTER 1

Max Planck

Introduction

1.1 Motivation

"One more thing... "

With this dramatic introduction Steve Jobs, founder and long-year CEO of Apple Inc., announced a new product in 2007 which rolled up the use of the internet and opened it up for mobile usage. The market entry of the first generation of the iPhone was the door opener to ground the former very cloudy ideas of Location-Based-Services (LBS) and their entry card for the mass market. With the high market penetration of smartphones soon after the introduction of the first iPhone the drive for higher integration regarding functionality and the implementation of new architectures exploiting the synergies of those functionality sped up. Two main factors, namely short lifetime on battery supply and the market's call for high speed performance pushed the development of new component generations.

Besides other sensors like gyroscope or camera, one of the central sensor functionality integrated in every smartphone available, is the ability to gain position information via a Global Navigation Satellite System (GNSS) receiver. While first generation smartphones relied on well known Acquisition and Tracking Receiver (ATR) chipsets soon the first assisted Navigational Satellite Timing and Ranging-Global Positioning System (NAVSTAR GPS) receivers were incorporated. The integration of data communication and positioning was again driven by the need for improved power consumption as well as fast operation, hence a fast position fix.

The first steps towards radio based navigation systems were already made with the upcoming of radio communication at the beginning of the last century overcoming the main drawback of former navigation solutions, the missing availability under bad environment conditions. One of the first navigation systems, built up during World War II and still being in operation, was the LORAN system, especially designed for the positioning of vessels on the ocean. Due to its system design, relying on Line-of-Sight (LOS) paths, the

LORAN systems still suffer from two main disadvantages. First a great amount of ground based transmitter stations must be available to cover earth's whole surface. The curvature of earth's surface limits the operating distance to approximately 1000 km. A lot of transmitting beacons were therefore built up to cover most areas of the world. Nevertheless the system was not available worldwide, especially not on parts of the ocean far away from bigger landmasses e.g. in the pacific region. Second precision of the positioning solution was limited to several tens of meters which is insufficient for navigation on land, especially in urban environments.

A way out of that dilemma was the development of satellite based transmitting beacons. Initially serving only military purposes, such as navigation possibilities for military vessels or missile guidance, the way into mass market was paved by the deactivation of the artificial degradation of the freely available Coarse Acquisition Code (C/A code) navigation signal of the NAVSTAR GPS in the year 2000 [9]. Since then a variety of civil applications have been developed based on navigation receivers, too. Nowadays toll- and logistic systems as well as personal navigation especially using mobile devices rely on GNSS receivers and have become part of our everyday life.

The demand of low power consumption is contrasted by the need for fast and reliable, and especially in case of navigation receivers, accurate functionality. Speed and reliability at first glance always come together with enhanced complexity which again leads to higher power consumption. Speaking in terms of navigation solutions a small Time-to-First-Fix (TTFF) is desired as well as highest accuracy even in bad signal reception environments. To meet the first demand new architectures besides the common ATR architecture have to be developed and integrated. The second demand can be met by the integration of second generation GNSSs.

The Galileo constellation doubles the number of satellites available and therefore gives a better chance of an undisturbed LOS to at least four satellites. However as referred to later in detail the integration of the second generation GNSSs arises new challenges to the receiver implementation which have to be taken into account in the development of new architectures.

1.2   Goals and Organization of this Work

The opposing demands of integrating second generation GNSSs as well as the need for low-power but still fast implementations require the development of new architectures for GNSS receivers. It will be demonstrated that the enhanced GNSS receiver architecture proposed, bridges the gap between these conflicting requirements. For the first time an assisted GNSS receiver solution is deeply coupled to a Long-Wave (LW) time code radio receiver using the synergies offered. Main advancement is the reuse and shrinking of major parts of common GNSS architectures for the calculation of the solution in time and space without deterioration of positioning solution accuracy.

The first assisted GNSS receiver architectures date back to the turn of the millenium [10]. These ideas gain superiority by cutting the TTFF using an alternative path for gathering the navigation data (satellite ephemeris data) and an open loop receiver architecture at the cost of increasing signal processing effort, especially in the digital domain. The signal processing complexity still arises from the absence of knowledge about signal reception time.

Although approaches have been proposed incorporating external sources for time information especially via an attached cellular network connection [11], those implementations still suffer from numerous restrictions. Without making claim to be complete, those are:

- Availability of a cellular network link.
- Precision of the time information.
- Power consumption due to query for up-to-date time information.

It will be demonstrated that the disadvantages of assisted as well as non-assisted architectures incorporating time information can be overcome. Mainly the massive computational power needed for the calculation of the code phases for the GNSS signals is reduced. Especially when implementing architectures similar to the single-shot receiver [12], which founds the basis for the developed architecture and serves as reference, advantages in calculation effort become evident.

Therefore after the introduction, chapter 2 gives an overview over the two most relevant existing global navigation satellite systems and their mode of operation, namely the NAVSTAR GPS and the Galileo System. Although subsequent parts of this work focus on positioning using NAVSTAR GPS signals, Galileo and its interoperability to NAVSTAR GPS and accordingly the integration into the newly developed architecture will be highlighted in some excursions, too. The focus was set on the integration of NAVSTAR GPS signals due to the unavailability of Galileo signals during the work underlying this thesis.

Mode of operation of the common ATR architecture will be highlighted together with its drawbacks, in chapter 3 in order to give understanding of the improvements which the new developed architecture will provide. Additionally, algorithms and operations described in this chapter are adopted for the newly developed architecture where applicable. Also a common assisted architecture, namely the single-shot receiver is introduced to complete the overview over state-of-the-art GNSS receiver architectures.

Principle and characteristics of long range time transfer as well as time code transmission are presented in chapter 4 with focus on the German DCF77 transmitter.

The following chapter 5 describes in detail the newly developed architecture. Focus is laid on the integration of assistance time information into the positioning solution. Benefits of the proposed architecture will be discussed in detail.

Results of a complex system simulation and implementation details for a demonstrator system will be given in chapter 6, proving functionality and benefits already pointed out in the preceding chapters.

The thesis is concluded in chapter 7 with a summary of this thesis and an outlook on future work.

CHAPTER 2

Global Navigation Satellite Systems

This chapter will give an overview over important Global Navigation Satellite Systems and their fundamental mode of operation. For NAVSTAR GPS as an example an in depth look into the algorithm of signal generation and positioning will be given and the differences to Galileo will be highlighted. Structure of this chapter and content, unless otherwise indicated, are based on the contents given by Kaplan et al. [13]. This work deals with an architecture aimed at the commercial market, hence it will not be qualified to decrypt the NAVSTAR GPS Y-encrypted Precise Code (P/Y code) or make use of the encrypted Galileo signals. Therefore only the relevant parts of the NAVSTAR GPS and Galileo Signal-in-Space (SIS) generation and signal transmission are investigated. For further information on the non-free of royalty or military encrypted codes and transmission frequencies besides the L1/E1 frequency bands the reader is referred to the literature [13], [14].

2.1   Principle of Satellite Navigation

2.1.1   Positioning using GNSS Space Vehicles

Positioning using GNSS Space Vehicles is based on a Time-of-Arrival (TOA) approach. Assuming precise knowledge about the position of the satellite, propagation speed of the signal (speed of light) and transmission as well as reception time, distances d between certain points in space and the transmitting beacon can easily be derived from equation (2.1). The parameters t_{send} and $t_{receive}$ represent transmission and reception time, respectively, and c the speed of light in vacuum.

$$d = \frac{t_{send} - t_{receive}}{c} \tag{2.1}$$

In two dimensions the distance d corresponds to a circle, in 3D to a sphere around the satellite. All valid solutions to equation (2.1) or the perimeter of the circle or sphere represent possible user positions based on a unique distance measurement. Using two transmission beacons, arranged in 2D space, reduce the number of valid results of the positioning solution to the number of two, P1 and P2, as illustrated in Figure 2.1.

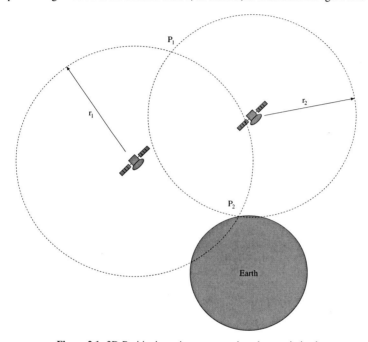

Figure 2.1: 2D-Positioning using two space based transmission beacons

This method can easily be expanded to three dimensions by adding a third transmitting beacon. Again two valid solutions P1 and P2 for the position of the user can be determined (see Figure 2.2).

The user position can then be computed using distances between satellite j and the user position r_j calculated by equation (2.2), calculating the absolute value of the difference between s_j, the position of the satellite j, and the users position u in Cartesian coordinates.

$$r_j = \|s_j - u\|$$
$$j = [1\ 3] \tag{2.2}$$

Precise knowledge about transmission and reception time is indispensable to calculate a satisfying positioning solution. The former is derived from highly precise time sources, atomic clocks and frequency normals on board each Space Vehicle (SV), guaranteeing

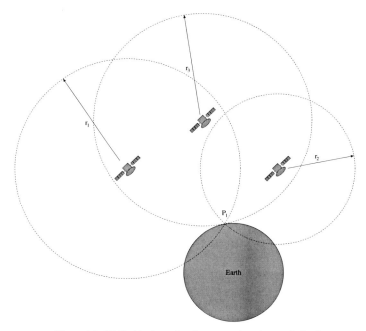

Figure 2.2: 3D-Positioning using three space based transmission beacons

highly precise time stamps which are incorporated in the satellite signal and transferred to the user. The latter, hence reception time, must be measured in the receiving device. Again a measurement with sufficient precision can be performed using atomic clocks tightly coupled to the used time reference namely Coordinated Universal Time (UTC). Unfortunately the integration of such high precise time sources is not trivial for mobile devices. Therefore the reception time mark cannot be measured with adequate precision and consequently the resulting uncertainty in the Time-of-Flight (TOF) measurement has to be taken into account when solving the system of equations for the positioning solution. An error term has to be added to equation (2.2) resulting in a fourth unknown in the system of equations (2.3).

$$
\begin{aligned}
r_j &= \|s_j - u\| + c\, t_{error} \\
j &= [1\ 4]
\end{aligned}
\tag{2.3}
$$

The fourth unknown t_{error}, besides user position coordinates u_x, u_y and u_z, introduced to the system of equations (2.3), results in an uncertainty in the range between user and satellite. The additional error term added to all equations in (2.3) introduces a new nomenclature for r_j. These are called pseudo-ranges, since the measured ranges to the satellites are inaccurate. To solve the resulting system of equations a fourth equation is needed which can be gathered from the measurement of the pseudorange to a fourth satellite.

Expanding equations (2.3) result in the system of equations (2.4).

$$
\begin{aligned}
r_1 &= \sqrt{(x_1 - x_u)^2 + (y_1 - y_u)^2 + (z_1 - z_u)^2} + c\, t_{error} \\
r_2 &= \sqrt{(x_2 - x_u)^2 + (y_2 - y_u)^2 + (z_2 - z_u)^2} + c\, t_{error} \\
r_3 &= \sqrt{(x_3 - x_u)^2 + (y_3 - y_u)^2 + (z_3 - z_u)^2} + c\, t_{error} \\
r_4 &= \sqrt{(x_4 - x_u)^2 + (y_4 - y_u)^2 + (z_4 - z_u)^2} + c\, t_{error}
\end{aligned}
\tag{2.4}
$$

This nonlinear system of equations can now be solved by different techniques either closed form solutions [15] [16] or iterative approaches [17] [13].

Still some questions regarding the variables in equation (2.4) have to be paid attention to. The most important, how coordinates of user and satellite positions are defined and how to obtain correct satellite positions at a certain time are answered in the following subsections.

2.1.2   Reference Coordinate Systems

The SV's Cartesian coordinates x_j, y_j, z_j as well as the user position coordinates x_u, y_u, z_u in equation (2.4) have to use a common base. The validity of the positioning solution is guaranteed by definition of a reference coordinate system. For NAVSTAR GPS two such coordinate systems are important depending on the actual purpose of usage, together with World Geodetic System 1984 (WGS-84), the standard worldwide chart datum :

- The Earth-Centered Inertial (ECI) coordinate system, which allows easy description of satellite orbits, mostly used to determine the SV orbits and calculate ephemeris data in the ground segment.

- The Earth-Centered-Earth-Fixed (ECEF) coordinate system which makes it easy to derive longitude, latitude and height from pseudorange measurements.

Calculations that are done using coordinates in the ECI coordinate system are mainly done by the control stations resulting in constellation and correction parameters for the orbits of the SVs described in the following section 2.1.3. For calculating the positioning solution the data is transferred to the ECEF coordinate system. This allows simple utilization of the correction and ephemeris data in the calculation of the SVs positions.

Although there are several mainly historically evolved chart data, the WGS-84 is the official worldwide standard model of the earth and output from the positioning solution is always converted to WGS-84 at first. In contrast to other chart data WGS-84 considers irregularities in the surface of the earth, its irregular density distribution and difference from a perfect spherical shape. These boundary conditions have major influence on the model for the SV motion and therefore have to be taken into account in the calculation of the SVs as well as the user position.

2.1.3 Satellite Orbit and Propagation Error Correction Parameters

Calculating the distances r_j in equation (2.4) between the SV and the user still two important factors remain after the definition of the reference coordinate system. Besides precise TOF measurement, which will be investigated later, it is highly critical to gather precise knowledge about the SV's orbits.

At first glance the SV orbit can be considered as a circle centered around the midpoint of the earth. This assumption only evaluates true for a perfect spherical shape and uniform mass distribution of earth. But as already pointed out in chapter 2.1.2 earth's surface shows a lot of variations from circularity and consequently a non uniform gravitational force. Since earth's gravity adds the dominant part of the acceleration to the satellite it mainly determines the SV's orbit.

Using a classic two-body motion approach for the motion of the satellite, the orbit of the SV can be described by the differential equation (2.5):

$$\frac{d^2\mathbf{r}}{dt^2} = \nabla V + \mathbf{a}_d \qquad (2.5)$$

with the initial position and velocity vectors $\mathbf{r}_0$ and $\mathbf{v}_0$ at the reference time t_0 as boundary condition.

The gravitational potential V in equation (2.5) models the force exerting for two-body motion approach while the acceleration correction parameter $\mathbf{a}_d$ combines all other disturbances like solar pressure, gravity influence from the sun and the moon, orbital maneuvers, Earth's tidal variations and others. The two-body motion equation, which ignores the disturbances described by $\mathbf{a}_d$ in equation (2.5), can analytically be solved resulting in six constants of integration. Regarding the full motion equation (2.5) the solution can only be calculated by numerical integration. Again the constants defined for the two-body motion problem can be used to characterize the orbit of the SV, too. But in contrast to the solution for the two-body problem the integrals are no longer constant and additionally only valid together with an initial position and velocity vector. These time-variable integrals, resulting from the solution of the full motion equation, therefore require the additional definition of a validity interval and a change rate for the integrals.

NAVSTAR GPS uses five of the six common Keplerian orbital elements together with the mean anomaly at epoch M_0 to describe the SV's orbit. While the first three Keplerian elements a, e, M_0 define the shape, the second three i, Ω and ω describe the orientation of the orbit in ECEF coordinates. The description of the elements is given in Table 2.1.

The 6th common Keplerian element, the time of perigee τ, is normally given as the third parameter of orbit shape for a two body motion solution. As already mentioned above the integrals become time variant if the full motion equation is calculated. Therefore it is preferable to specify the orbit with time dependent Keplerian elements. For this reason the common Keplerian element τ is replaced by the equivalent mean anomaly at epoch M_0.

The Keplerian elements defining the shape of the orbit, except the mean anomaly at epoch t_0, can be illustrated geometrically. The last parameter, the mean anomaly M_0 at epoch t_0, instead cannot be plotted since it has no geometric equivalent. Nevertheless it can be

Keplerian element	Description
a	Semimajor axis of the ellipse
e	Eccentricity of the ellipse
M_0	Mean anomaly in epoch
i	Inclination of orbit
Ω	Longitude of ascending node
ω	Argument of perigee

Table 2.1: Definition of Keplerian elements for NAVSTAR GPS satellite orbits

calculated from the mean anomaly M. This parameter in turn results from the conversion of the the true anomaly ν according to equations (2.6). ν again is illustrated (see Figure 2.3).

$$M - M_0 = \sqrt{\frac{\mu}{a^3}}(t - t_0)$$
$$M = E - e \sin E \tag{2.6}$$
$$E = 2 \arctan \left[\sqrt{\frac{1-e}{1+e}} \tan \left(\frac{1}{2}\nu \right) \right]$$

An illustration of the already mentioned Keplerian elements i, Ω, ω defining the orientation of the SV's orbit and the mean anomaly at epoch M_0 is given in Fig. 2.3. The reference direction is equal to the x-axis of the ECEF coordinate system defined in 2.1.2.

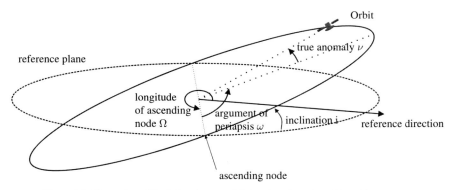

Figure 2.3: Illustration of Keplerian elements defining the orientation of the orbit

The complete ephemeris or orbit prediction data structure used for the calculation of GNSS SV's position is given in Table 2.2 with respect to the nomenclature used in [7]. The time variant Keplerian elements, whose validity expires after a short period, are enhanced by correction parameters, which allow a sufficiently accurate prediction of the

Ephemeris element	Description
t_{0e}	Reference time of ephemeris
$\sqrt{a}$	Squareroot of semimajor axis of the ellipse
e	Eccentricity of the ellipse
i_0	Inclination angle at time t_{0e}
Ω_0	Longitude of ascending node (at weekly epoch)
ω	Argument of perigee at time t_{0e}
M_0	Mean anomaly in epoch at time t_{0e}
$\frac{di}{dt}$	Rate of change of inclination angle
$\dot{\Omega}$	Rate of change of longitude of the ascending node
δn	Mean motion correction
C_{uc}	Amplitude of cosine correction to argument of latitude
C_{us}	Amplitude of sine correction to argument of latitude
C_{rc}	Amplitude of cosine correction to orbital radius
C_{rs}	Amplitude of sine correction to orbital radius
C_{ic}	Amplitude of cosine correction to inclination angle
C_{is}	Amplitude of sine correction to inclination angle

Table 2.2: Definition of ephemeris elements for NAVSTAR GPS satellite orbits

satellite's motion even for a longer period of time [13], [14].

Besides the strictly geometric parameters describing the orbit, secondary variables have to be considered when calculating the positioning solution, too. Namely, delays introduced by two atmospheric layers, the ionosphere and the troposphere have to be accounted for.

The influence of the ionosphere, which is the greatest single error source in the propagation of the SV's signals [18], can be described by the Klobuchar model [19]. Although it only covers 50 % of the ionospheric propagation error it is still sufficient for single frequency NAVSTAR GPS applications [20]. The model allows the calculation of the ionospheric delay based on the variables given in Table 2.3.

Element	Description
α_n	Coefficients of a cubic equation representing the amplitude of the vertical delay (for the n-th satellite)
β_n	Coefficients of a cubic equation representing the period of the model (for the n-th satellite)
E	Elevation angle between the user and satellite (semi-circles)
A	Azimuth angle, measured clockwise, between user and satellite from the true North (semi-circles)
ϕ_u	Geodetic latitude of users position (semi-circles) in WGS-84 coordinates
λ_u	Geodetic longitude of users position (semi-circles) in WGS-84 coordinates
gps time	NAVSTAR GPS system time

Table 2.3: Definition of ionospheric model parameters for the propagation of the L1-NAVSTAR GPS signals

These variables are then inserted into equations (2.7).

$$
t_{iono} = \begin{cases} F\left[5,0\ 10^{-9} + AMP\left(1 - \frac{x^2}{2} + \frac{x^4}{4}\right)\right], & |x| < 1,57\,\text{s} \\ F\left(5,0\ 10^{-9}\right), & |x| \geq 1,57\,\text{s} \end{cases}
$$

$$
AMP = \begin{cases} \sum_{n=0}^{3} \alpha_n \phi_m^n, & |AMP| \geq 0\,\text{s} \\ 0, & |AMP| < 0\,\text{s} \end{cases}
$$

$$
PER = \begin{cases} \sum_{n=0}^{3} \beta_n \phi_m^n, & |PER| \geq 72\,\text{s} \\ 72, & |PER| < 72\,\text{s} \end{cases}
$$

$$
x = \frac{2\pi(t - 50400)}{PER}
$$

$$
F = 1,0 + 16[0,53 - E]^3
$$

(2.7)

The geomagnetic latitude of earth's projection of the ionospheric intersection point ϕ_m and the local time t are either derived from the users position data or from the NAVSTAR GPS system time, respectively [7]. Obviously correction of ionospheric influence is dependent on position and time information. Both are not available to the device, since these are the fundamental deliverables of the device functions. It will be shown how to solve this paradox by iterative means of computation or integration of assistance data.

Based on [21] and [22] the computation of signal delays based on tropospheric influences, mainly humidity, is dominated by the influence of the elevation angle of the satellite based on the geodetic height and latitude of the user position.

The SV's actual position can then be computed solving orbit equations (2.5) together with the ephemeris data and actual time information according to equations (2.8).

$$x_s = x_p \, \cos(\Omega_k) - y_p \, \cos(i_k \, \sin(\Omega_k)$$
$$y_s = x_p \, \sin(\Omega_k) + y_p \, \cos(i_k \, \cos(\Omega_k)$$
$$z_s = y_p \, \sin(i_k)$$

$$x_p = r_k \, \cos(u_k)$$
$$y_p = r_k \, \sin(u_k)$$

$$\Omega_k = \Omega_0 + (\dot{\Omega} - \dot{\Omega}_e)(t_k) - \dot{\Omega}_e t_{0e}$$
$$i_k = i_0 + (\frac{di}{dt})t_k$$
$$r_k = a(1 - e \cos E_k)\delta r_k$$
$$u_k = \phi_k + \delta\phi_k$$

$$\delta\phi_k = C_{us} \sin(2\phi_k) + C_{uk} \sin(2\phi_k)$$
$$\delta i_k = C_{is} \sin(2\phi_k) + C_{ic} \sin(2\phi_k)$$
$$\delta r_k = C_{rs} \sin(2\phi_k) + C_{rc} \sin(2\phi_k)$$

$$\phi_k = \nu_k + \omega$$

$$\sin \nu_k = \frac{\sqrt{1 - e^2} \sin E_k}{1 - e \, \cos E_k}$$
$$\cos \nu_k = \frac{\cos E_k - e}{1 - e \, \cos E_k}$$

$$M_k = E_k - e \, \sin E_k$$
$$M_k = M_0 + n(t_k)$$
$$t_k = t - t_{0e}$$
$$n = \sqrt{\frac{\mu}{a^3}} + \Delta n$$

(2.8)

2.2 NAVSTAR GPS

The development of the NAVSTAR GPS system was started on the early 1970s. The goal was to develop a reliable and precise positioning system which overcomes the issues of its predecessors. Especially worldwide availability at 24/7, which was not guaranteed by e.g. the Navy Navigation Satellite System (TRANSIT), was a goal during the development. Although the design of NAVSTAR GPS was initiated by the armed forces of the United States, namely the United States Air Force, it soon became clear that it would be available for civil users, too. The first satellites were launched in 1978 and full constellation was reached in 1989.

The massive spread of GNSS applications in digital life was pushed by the deactivation of the Selective Availability (SA), an artificial degradation of the time information, in the year 2000 [9]. Not until then reasonable accuracy was possible using NAVSTAR GPS's open service C/A code.

2.2.1   NAVSTAR GPS System Segments

The NAVSTAR GPS system is composed of three segments, the user-, control- and space segment.

The space segment includes all satellites of the NAVSTAR GPS. In order to ensure the coverage of the whole surface of the earth with at least 3 satellites in view at all times a constellation with 6 orbit planes each equipped with a minimum of 4 satellites in equivalent distances to each other was chosen. Each orbit plane has an inclination of 55 ° above the equator. In comparison to each other the orbital planes differ 60 ° in the equatorial plane. All satellites move at an altitude of approximately 20200 km [23]. The constellation is illustrated in Fig. 2.4.

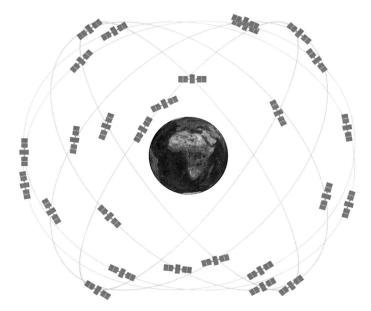

Figure 2.4: Constellation of NAVSTAR GPS SV [1]

The control segment is composed of the Master Control Station (MCS) located at Shriever Airforce Base (AFB) in Colorado, USA and 10 monitor stations around the world. Their locations can be derived from Figure 2.5. The locations have been carefully chosen so that each SV can be tracked by at least two stations simultaneously [13]. This is essential in order to measure deviations of the satellite's calculated orbits. These deviations are integrated into the navigation message by the master control station. The payload data, including the navigation message as described in 2.1.3, will be described in section 2.2.2. The corrected data calibrates the positioning solution removing errors due to the time variant integrals. The users can therefore be provided with updated orbit information by the master control station by an updated navigation message.

Figure 2.5: Location of NAVSTAR GPS monitor stations

The users themselves or mainly their navigation capable devices span the user segment.

2.2.2   Signal-in-Space Generation and Characteristics

The NAVSTAR GPS SIS has to fulfill multiple requirements. First, despite the payload data, precise information about time of transmission has to be transferred to the user segment, too. Second the user segment must be able to differentiate signals from different SVs from each other. At last the signal design has to overcome the problem of limited signal transmission power at the SVs, nevertheless allowing the reception, demodulation and decoding of the transmitted information in the user segment, even under bad reception conditions.

To fulfill these demands a Code Division Multiple Access (CDMA) spreading technique was chosen. Besides the differentiation of the NAVSTAR GPS SVs signals the nature of the spreading codes allows transfer of precise time information to the user segment. While the orthogonality between the Gold Codes [24], uniquely transmitted by each satellite, allows identification of the SVs, the time transfer is realized by the incoming code phase of the satellite signal. The spreading technique is also useful to overcome the problem of low transmission power. This is highlighted in section 3.1.

The generation of the Pseudo-Random Noise Sequences forming the CDMA codes is shown in Fig. 2.6.

The specific Pseudo-Random Noise Sequence (PRN) is generated from two M-sequences

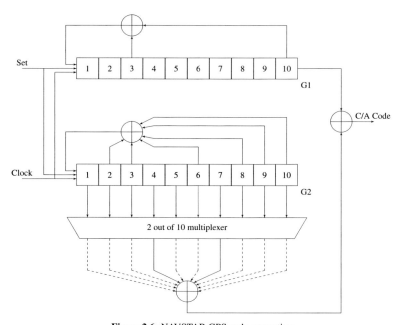

Figure 2.6: NAVSTAR GPS code generation

[25] which are derived from two Linear Feedback Shift Registers (LFSRs). The combination of the M-sequences via an Exclusive-Or Gate (XOR gate)-function form Gold-Codes [24] which are especially qualified for the usage in GNSS due to their excellent Auto Correlation Function (ACF) and Cross Correlation Function (CCF) properties. The instruction for generating the first sequence G1 is equation (2.9)

$$G_1 = x^{10} + x^3 \tag{2.9}$$

while the generation of the second sequence G2 follows equation (2.10)

$$G_2 = x^{10} + x^9 + x^8 + x^6 + x^3 + x^2 \tag{2.10}$$

While G1 is directly used for the generation of the C/A code, specific taps of the G2 register are fed into an XOR gate for generating the second sequence for the C/A code generation. By using different taps of the G2 register it is possible to produce different code sequences. The specific taps for the active NAVSTAR GPS SVs are shown in Table 2.4 [14]. A XOR gate combines G1 and the resulting bit sequence from the G2 register taps to the C/A code.

LFSRs of length m allow the generation of 2^m-1 long sequences until repetition of the

SV number	G2 Taps	SV number	G2 Taps
1	$2 \oplus 6$	17	$1 \oplus 4$
2	$3 \oplus 7$	18	$2 \oplus 5$
3	$4 \oplus 8$	19	$3 \oplus 6$
4	$5 \oplus 9$	20	$4 \oplus 7$
5	$1 \oplus 9$	21	$5 \oplus 8$
6	$2 \oplus 10$	22	$6 \oplus 9$
7	$1 \oplus 8$	23	$1 \oplus 3$
8	$2 \oplus 9$	24	$4 \oplus 6$
9	$3 \oplus 10$	25	$5 \oplus 7$
10	$2 \oplus 3$	26	$6 \oplus 8$
11	$3 \oplus 4$	27	$7 \oplus 9$
12	$5 \oplus 6$	28	$8 \oplus 10$
13	$6 \oplus 7$	29	$1 \oplus 6$
14	$7 \oplus 8$	30	$2 \oplus 7$
15	$8 \oplus 9$	31	$3 \oplus 8$
16	$9 \oplus 10$	32	$4 \oplus 9$

Table 2.4: Taps for SVs separation [7]

code sequence [24]. In case of the NAVSTAR GPS C/A code signal the 10 bit long LFSRs generating G1 and G2 result in PRNs of 1023 chips length. In order to achieve a duration of 1 ms for the PRN sequences the LFSRs have to be clocked with 1.023 MHz. By setting the registers G1 and G2 every 1 ms the 50 Hz data clock frequency for the navigation message can easily be derived by an all 1's logic at the taps of the G2 register, generating a 1 kHz signal which is then divided by a divide-by-20 clock divider.

The 1.023 MHz clock, needed for the generation of the PRN sequences, is derived from a 10.23 MHz master clock which additionally is produced by the satellite's on board atomic clocks [7]. As shown in Fig. 2.7 the master clock is again used for the generation of the Radio-Frequency (RF) carrier frequency by feeding it into a frequency multiplier (multiplication by 154 in case of the L1-C/A code signal).

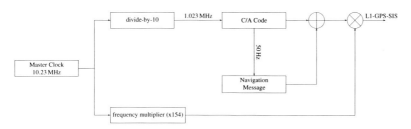

Figure 2.7: NAVSTAR GPS SIS generation

The navigation message consists of 37500 bit transmitted at 50 $\frac{\text{bits}}{\text{s}}$ and provides the user with up-to-date ephemeris data introduced in section 2.1.3 as well as clock error and atmospheric signal propagation data. The complete message is divided into 25 frames, each 1500 bit long. Every frame again is partitioned into 6 sub-frames, each consisting of ten words represented by 30 bit as shown in Fig. 2.9 [7].

Each subframe starts with the Telemetry Word (TLM) which contains a preamble to give the receiver the opportunity to find the starting point of each subframe. The Hand-Over-Word (HOW), which is word number two in each subframe, contains the number of epochs past since the start of last week and ensures "handover" to the encrypted P(Y)-code for qualified users.

Words three to ten of each subframe contain the payload data. Neither all payload data nor its arrangement in the navigation message is essential in the primary scope of this work. Therefore only the relevant parts are discussed in detail. An in depth look on all data and its organization is given in [7].

While subframes one to three contain SV-specific clock correction and the already discussed orbit parameters (see section 2.1.3), data in subframes four and five carry the almanach data.

Clock correction parameters have to be applied to the positioning solution despite the fact that the SVs are carrying several atomic clocks due to characteristics of the SV itself. These include deterministic effects like bias, drift or aging which have to be estimated.

Tables 2.5-2.8 show the organization and position of the ephemeris parameters included in the corresponding subframe.

Parameter	no. of bits	Description
Week no.	10	10 MSBs of NAVSTAR GPS week number, indicates the number of week epochs since 01.01.1980
satellite accuracy	4	User Range Accuracy [m]
satellite health	6	MSB defines health of navigation data ('0' navigation data is OK)
T_{GD}	8	Estimated Group Delay Differential
IODC	10	Issue of Data (Clock)
t_{oc}	16	Satellite Clock Correction Parameter
a_{f2}	8	Satellite Clock Correction Parameter
a_{f1}	16	Satellite Clock Correction Parameter
a_{f0}	22	Satellite Clock Correction Parameter

Table 2.5: Subframe one parameters: specific satellite clock and health data [7]

Subframes four and five are transmitted in 25 different versions, called pages, each containing different data.

All pages of subframes four and five start with TLM word and HOW. Word three contains the data ID and depending on the payload data in words four to ten either the PRN number of the corresponding SV or for all other pages the ID of the transmitting satellite.

Parameter	no. of bits	Description
IODE	8	Issue of Data (Ephemeris)
C_{rs}	16	Amplitude of the Sine Harmonic Correction Term to the Orbit Radius
Δn	16	Mean Motion Difference from Computed Value
M_0	32	Mean Anomaly at Reference Time
C_{uc}	16	Amplitude of the Cosine Harmonic Correction Term to the Argument of Latitude
e	32	Eccentricity
C_{us}	16	Amplitude of the Sine Harmonic Correction Term to the Argument of Latitude
$A^{\frac{1}{2}}$	32	Square Root of the Semi-Major Axis
t_{oe}	16	Reference Time Ephemeris
C_{ic}	16	Amplitude of the Cosine Harmonic Correction Term to the Angle of Inclination
$OMEGA_0$	32	Longitude of Ascending Node of Orbit Plane at Weekly Epoch
C_{is}	16	Amplitude of the Sine Harmonic Correction Term to the Angle of Inclination
i_0	32	Inclination Angle at Reference Time
C_{rc}	16	Amplitude of the Cosine Harmonic Correction Term to the Orbit Radius
ω	32	Argument of Perigee
OMEGADOT	24	Rate of Right Ascension
IDOT	14	Rate of Inclination Angle

Table 2.6: Subframe two and three parameters: specific satellite ephemeris parameters [7]

Page	Description
2,3,4,5,7,8,9,10	Almanac data for satellite 25 through 32 respectively
17	Special messages
18	Ionospheric and UTC data
25	Satellite configurations for 32 satellites
1,6,11,12,16,19,20,21,22,23,24	Reserved
13,14,15	Spares

Table 2.7: Subframe four pages: support data [7]

The almanac data encoded in pages two through five and seven through ten of subframe four and pages one through 24 of subframe five respectively contains a number of parameters listed in Table 2.9

It is obvious that comparing Tables 2.5 through 2.8 and Table 2.9 that the almanac con-

Page	Description
1 through 25 Week	Almanac data for satellite 1 through 24
25	Satellite health data for satellite 1 through 24, almanac reference time and almanac reference week number

Table 2.8: Subframe five pages: support data [7]

Parameter	no. of bits	Description
e	16	Issue of Data (Ephemeris)
t_{oa}	8	Amplitude of the Sine Harmonic Correction Term to the Orbit Radius
δi	16	Mean Motion Difference from Computed Value
OMEGADOT	16	Rate of Right Ascension
$A^{\frac{1}{2}}$	16	Square Root of the Semi-Major Axis
$OMEGA_0$	24	Longitude of Ascending Node of Orbit Plane at Weekly Epoch
ω	24	Argument of Perigee
M_0	24	Mean Anomaly at Reference Time
a_{f0}	11	Satellite Clock Correction Parameter
a_{f1}	11	Satellite Clock Correction Parameter

Table 2.9: Subframe five pages: support data [7]

sists of a subset of the clock and ephemeris parameters of each SV. Additionally, some parameters are truncated and therefore their precision is declined. The meaning of the truncation in precision of the almanac parameters become clear, when remembering that precise data can be decoded from the ephemeris in the signal of the specific SV itself, assuming it is in view, after synchronization onto its data signal. Looking back to Figure 2.7 the navigation message is modulated to the PRN code by an modulo-2 addition operation, generating inverse PRN sequences for binary '1's and '0's of the navigation message.

The resulting signal is modulated onto the carrier frequency of 1.57542 GHz. Later developments incorporate a second NAVSTAR GPS carrier frequency at 1.2276 GHz. Figure 2.8 gives an overview over available GNSSs, their carrier frequencies and an idea of the signal bandwidth.

2.3   Galileo

Based on the memorandum of understanding regarding interoperability between GPS and Galileo [26] [27] both satellite sytems are designed to work in the same frequency band yet at the same carrier frequency. Due to different modulation schemes, especially developed to fulfill the demand of interoperability, as well as modified non-analytical PRN

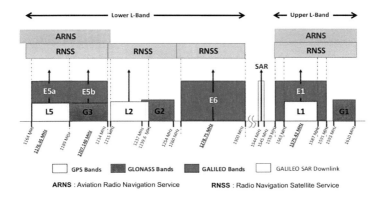

Figure 2.8: Frequency plan of existing GNSSs [2]

sequences it is possible to separate the signals from both GNSSs. Nevertheless there are a lot of similarities, why mainly the differences between both systems are carved out in the following.

2.3.1   Galileo System Segments

Similar to NAVSTAR GPS the Galileo system consists of the three system segments, the user-, control- and space segment. Again the goal is to ensure the coverage of the whole surface of the Earth with at least four satellites in view at all times. Galileo is planned as a Walker constellation [28] of 30 satellites, as illustrated in Fig. 2.10, with three orbital planes inclined 56° above the equator. The altitude of the orbital planes is 23222 km [29].

The Galileo control segment is splitted into the Ground Control Segment (GCS) and the Ground Mission Segment (GMS). The GCS is responsible for the functionality of the Galileo system, especially monitoring and adjusting the constellation. Therefore five control sites around the world will be set up. Main facilities are the master control stations at Oberpfaffenhofen, Germany and Fucino, Italy. Regarding the GMS, which is responsible for the correct navigation signals, the master control stations are augmented by 40 monitor stations around the world, analog to those augmenting the NAVSTAR GPS system.

Again the user himself or mainly the users navigation capable device can be described as the user segment.

2.3.2   Signal in Space Characteristics

Similar to NAVSTAR GPS, the Galileo SIS will be composed of different signal components. Multiple carrier frequencies transport different ranging codes (encrypted as well as unencrypted) to the user segment. Additionally, the navigation data payload in differ-

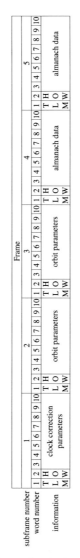

Figure 2.9: Format of navigation message frame

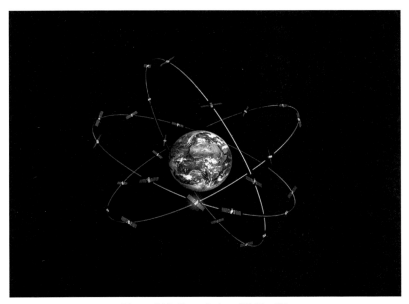

Figure 2.10: Constellation of Galileo [3]

ent precisions will be transmitted, carrying information about the constellation as well as distance information, according to Table 2.10.

Signal	Carrier Frequency	Services
L1	1.575420 GHz	OS, CS, SoL, PRS
E6	1.278750 GHz	CS, PRS
E5	1.191795 GHz	OS, CS, SoL

Table 2.10: Galileo signals and corresponding carrier frequencies

Again similar to NAVSTAR GPS, each individual service will have unique restrictions and will be transmitted on multiple carrier frequencies. In contrast to the two services of NAVSTAR GPS, C/A code and P/Y code, Galileo will provide more services, each at least transmitted on two carrier frequencies (see column three of table 2.10). These services include:

- OS: Open Service, free of royalty, similar to NAVSTAR GPS C/A code

- CS: Commercial Service, liable to pay costs, encrypted, higher data rate of navigation message for improved accuracy, Quality-of-Service (QoS) information

- PRS: Public Regulated Service, encrypted, available only to public authorities, especially hardened against jamming and spoofing

- SoL: Safety-of-Life Service, safety-critical service, encrypted, especially for air- and vehicular appliances with special qualified receivers

Since only the L1-Galileo Open Service (OS) is of interest here it will be highlighted and especially its differences and similarities to the NAVSTAR GPS C/A code signal. Other signal components and carrier frequencies are described in depth in [30].

Based on the memorandum of understanding between the European Union (EU) and the United States of America (USA) [26] co-existence of both systems in the same frequency band even usage of the same carrier frequency of 1.57542 GHz is possible due to special precautions taken in the generation of the Galileo signal. A block diagram of the Galileo SIS generating circuit is given in Figure 2.11.

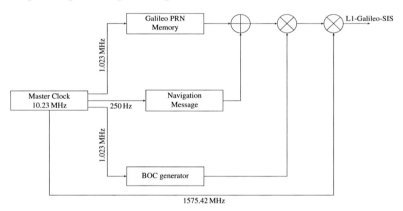

Figure 2.11: Block diagram of Galileo signal generation

At first glance there are two main differences in the generation of the Galileo signal compared to the NAVSTAR GPS C/A code signal. Besides the higher clock frequency for the navigation data of $250\ \frac{bits}{s}$, the modulation with a rectangular subcarrier prior to the mixing process to the carrier frequency is eye-catching.

While the NAVSTAR GPS C/A code has one main lobe in its spectrum at the carrier frequency, the Galileo signal is spread into two side lobes by the mixing process with the subcarrier. The resulting spectrum corresponds to the two sidelobes, slightly shifted besides the carrier frequency of 1.57542 GHz. Depending on the design of the rectangular subcarrier, peak-to-peak distance and occupied bandwidth of the side lobes can be adjusted. This modulation scheme is called Binary Offset Carrier (BOC) modulation [31]. The BOC modulation is always denoted BOC(n,m), where n defines the distance of the maximum of the side lobe from the nominal carrier frequency in multiples of the chipping rate, while m stands for the factor the side lobe is spread in comparison to an unmodulated signal.

For the Galileo L1-OS a BOC(1,1) [30] modulation was chosen. From Figure 2.12 it can easily be seen that BOC(1,1) modulation allows Galileo signals in the same band with the NAVSTAR GPS L1-signal without interfering with each other.

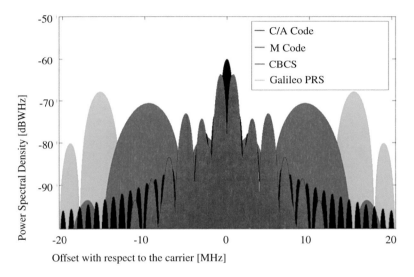

Figure 2.12: Spectra of Galileo E1 OS and NAVSTAR GPS C/A code signals [4]

Another significant difference between both GNSS is located in the PRN ranging codes. While NAVSTAR GPS relies on Gold codes generated from LFSRs, the Galileo L1-OS PRNs do not follow analytic equations and can therefore not be generated by logical circuits. Instead these so called memory-codes have to be hard-coded into memory of the Galileo SIS generating unit. The codes are enlisted in [30].

As mentioned above a navigation message as part of the Galileo signal transports information about the position of the actual SV as well as about the constellation to the user segment. Galileo defines different navigation messages for its services which differ in the accuracy of the transported information. The L1-OS signal carries the F/NAV message (additionally the I/NAV and C/NAV messages exist which are not of interest here [30]). This navigation message is organized in sequential frames, each 600 s long and containing 12 subframes. These subframes each 50 s long are divided into five pages of 10 s length, each.

The F/NAV pages present data similar to the NAVSTAR GPS subframes. Each page starts with a synchronization pattern, formed by a fixed bit sequence. This preamble is followed by the F/NAV symbols which transport information about the page type, the navigation data, Cyclic Redundancy Check (CRC) information and tail bits.

The frame is built up from subframes alternately incorporating pages of type five or six as last page. Depending on odd or even frame number, the last pages of the frame are switched. Both sub-frame types are illustrated in tables 2.11 and 2.12. The parameter k is set by the Galileo system for each SV, while l stands for the actual frame number.

Page Type	Page Content
1	SVID, clock correction, SISA, Ionospheric correction, BGD, Signal health status, GST and Data validity status
2	Ephemeris (1/3) and GST
3	Ephemeris (2/3) and GST
4	Ephemeris (3/3), GST-UTC conversion, GST-gps Conversion and TOW
5	Almanac for satellite k+(l-1) and almanac for satellite (k+l) part 1

Table 2.11: Odd-numbered subframe of Galileo F/NAV message

Page Type	Page Content
1	SVID, clock correction, SISA, Ionospheric correction, BGD, Signal health status, GST and Data validity status
2	Ephemeris (1/3) and GST
3	Ephemeris (2/3) and GST
4	Ephemeris (3/3), GST-UTC conversion, GST-GPS Conversion and TOW
6	Almanac for satellite (k+l) part 2 and almanac for satellite (k+l+1)

Table 2.12: Even-numbered subframe of Galileo F/NAV message

Clock correction parameters, ephemeris and almanac data are transported by the F/NAV message, although arranged in a different order, including the same parameters given in 2.2.2. The navigation message only differs in the definition of Galileo System Time (GST). The freezing point of GST is 00:00:00 on August the 22nd, 1999 with 13 leap seconds ahead of UTC. Taking the different focal point in time into account the calculation of the SV's position is equivalent for Galileo and NAVSTAR GPS.

CHAPTER 3

State-of-the-Art GNSS Receiver Architectures

In this chapter, two common GNSS receiver architectures are presented. In order to understand the supremacy of Assisted Global Navigation Satellite System (AGNSS) receiver concepts, first a non-assisted architecture, namely the Acquisition and Tracking Receiver is examined. Second an AGNSS architecture, the single-shot receiver, is introduced. The concept of gathering assistance data through a wireless connection and its influence on the positioning solution will be highlighted as well as key figures of the RF Frontend and the baseband architecture. Finally, an insight will be given on the remaining drawbacks of the presented architectures.

3.1 Acquisition and Tracking Receiver Architecture

The ATR architecture, shown in Fig. 3.1, represents a common GNSS receiver architecture. The partitioning into acquisition and tracking is based on the two fundamentally different modes of operation of the receiver. Turning on a navigation receiver, without any a priori knowledge about e.g. approximate position, time or navigation data, at first those pieces of information must be acquired and made available to the receiver. This information including constellation parameters as well as time information are essential in order to calculate a positioning solution which represents the end of the acquisition phase. Second during tracking phase, according to its name, only the actual code phase of the SVs in view is tracked and the positioning solution is constantly updated.

The difference in both modes of operation is situated in the way the data generated in the signal baseband processing is processed further. Prepending, the RF Frontend has to apply analog signal processing, mainly filtering and amplification, to the incoming signal. Several concepts for GNSS RF Frontends fulfilling the demands, have been proposed in literature [13] [32].

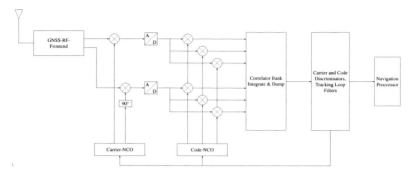

Figure 3.1: Block diagram of ATR architecture

The baseband signal processing is based on the Integrate-and-Dump (I&D) concept [13]. It uses the CCF between the received signal and a local generated PRN replica as a figure of merit to acquire and track the specific satellite signal as well as to decode the navigation message. First the acquisition part is required to determine the initial code phase of the satellite signal. Therefore the I&D receiver architecture integrates the results of the CCF between the incoming GNSS signal and a local generated replica for a specific satellite for a certain time. The result is compared to a threshold value. To raise the result of the CCF above the threshold value even under bad reception conditions (low signal strengths) integration time can be varied. If the integration result does not exceed the threshold value it is dumped and a new integration is started with a different initial shift applied to the local generated replica.

If the result in the present path of the Early-Present-Late (EPL) structure in Fig.3.1 exceeds the estimated threshold value the results of the adjacent correlation channels named "early" and "late" and in case of Galileo signals additionally "very early" and "very late" are examined. As soon as the EPL or Very Early-Early-Present-Late-Very Late (VEEPLVL), deliver results of the CCF as given in Figures 3.2(a) and 3.2(b) respectively, the initial code phase shift has been identified.

Once the correct code phase has been acquired the receiver is switched to tracking mode where the receiver starts to decode the navigation message. Remembering the definition of the NAVSTAR GPS or Galileo SIS, the navigation message is modulated on the SIS by either inverting the PRN sequence or not. This causes the CCF to deliver results of the same magnitude but different sign. Dependent on the sign of the CCF maximum now the decision whether a binary '1' or '0' was decoded is made [13].

During and after reception of the navigation message the loop structure of the architecture adjusts the code generator in Fig. 3.1 so that the global maximum of the CCF is constantly the result of the present path of the EPL structure [33]. The distance in chips, which the local replica of PRN sequences has to be shifted, corresponds to the offset between actual local time (thus local's Real-Time-Clock (RTC) time) and NAVSTAR GPS time. Together with transmission time and associated satellite position, both encoded in the navigation message (see section 2), the current distance to the satellite can be calculated. Consequently, the corresponding pseudorange, still defected by the inaccuracy of reception time measurement, can be determined. Having four parallel pseudorange measurements to four different satellites available, the position of the receiver can be computed according to the equations already presented in section 2.1.1.

The main drawback of the ATR architecture is a very long TTFF after a cold start of the positioning device. A cold start is defined as the power up of the navigation device without any information or estimation neither about local position and time, nor about the satellite constellation. Power is consumed permanently during reception of the navigation message and until the first position fix is calculated. A second possible scenario is the deterioration of the navigation message due to a long time since the last download of the data. A brute force search for every possible SV, each over all possible code phases and the Doppler drift range has to be started to identify the SVs in view. This can be done serially with low hardware effort which then requires a long time until the first valid position is calculated. Acceleration can be achieved by the use of parallel search structures. Both result in unacceptable high power consumption or long TTFF, either.

Both modes of operation, acquisition and tracking, are not energy efficient. Decoding the full navigation message relying one satellite signal nominally takes 12.5 min. During that time the whole receiver has to be powered, preventing loss of lock to the code phase of the initial acquired SV. Although the time until the full navigation message is available can be reduced by parallel decoding of several satellite signals, increased speed is again paid with increased power consumption.

In tracking mode, although the navigation message is already available, the ATR suffers from significant high power consumption. Again all parts of the receiver, except the redundant parts for identifying the initial code phase, has to be powered as long as the receiver is active, preventing the tracking loops from unlocking.

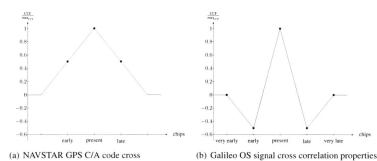

(a) NAVSTAR GPS C/A code cross correlation properties

(b) Galileo OS signal cross correlation properties

Figure 3.2: GNSS signal cross correlation properties

3.2   Single-Shot Receiver Architecture

The Single-Shot-Receiver architecture, firstly specified by Akopian [10], was one of the first assisted GNSS architectures introduced. Assisted receivers have access to information about an approximate local position and time as well as the navigation message through other sources than the satellite signal itself. In case of mobile devices the collection of this information can be realized through an attached cellular network connection. While direct sources for approximate time and navigation message exist (e.g. 3GPP assistance protocol [34]) the approximate position can be derived by the cellular network, too. The easiest way of determining an approximate position is to adopt the position of the base station the device is logged onto and reading the cell ID of the cellular network. Querying a database connecting cell ID and geographic position consequently delivers an inital position approximation.

The division of the positioning process into acquisition and tracking, as immanent to the ATR, is no longer applicable to assisted architectures. Acquisition in its common interpretation, hence the determination of satellites in view and the associated initial code phases are obsolete, since the SVs in view can be predicted by bringing approximate position, time and navigation message together. Consequently the cold-start scenario is no longer immanent to the architecture. Since the navigation message is already available, the tracking part of common architectures which is responsible for the decoding of the navigation message and constantly determining the current code phase is now obsolete in the common specification, too.

Consequently the carrier and code phase tracking loops can be unsewed and a loop free architecture is arising. Nevertheless, the I&D principle still represents the basis for the Single-Shot-Receiver (SSR) baseband signal processing architecture.

Comparing warm and hot start scenarios of assisted and non-assisted architectures, the TTFF do not differ significantly. Instead the big differences can be observed in the mean power consumption. While navigation in assisted mode is relying on interpolation between discrete positioning solutions, non-assisted architectures are dependent on seamless tracking of the satellite signals. Therefore assisted architectures gain their advantage by deactivating key components of the receiver in between the positioning points in time. Not only the analog RF Frontend, of the receiver but even the baseband signal processing can be powered down while the positioning solution is approximated by Kalman filtering in between. Applying this method to ATR receivers would cause the tracking loops to unlock and consequently consume time and power to reacquire the SVs's code phases.

The block diagram in Figure 3.3 gives an overview of the single-shot receiver architecture. Only the relevant parts, hence the assisted portions of the receiver, are illustrated. Of course a fallback to the common ATR must be possible in case no assistance data is available.

The main difference in the mode of operation, compared to the ATR, is situated in the baseband signal processing, while the RF Frontend can be reused without modification.

The result of the CCF between the received satellite signal and a local generated replica is used to derive the actual code phase of the satellite signal and hence the distance to the satellite. As already stated in section 3.1 this can be realized by an I&D architecture in the

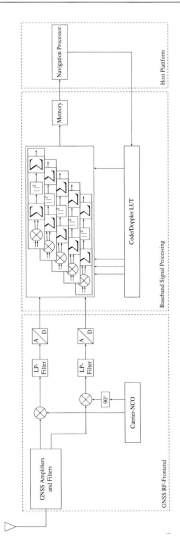

Figure 3.3: Block diagram of Single-Shot architecture

baseband signal processing, implementing three correlator channels in an EPL structure for NAVSTAR GPS or five channels in a VEEPLVL structure for Galileo, respectively. The basic concept of the correlator channel is picked up again by the Single-Shot-Receiver (SSR) baseband engine but the search window is widened from three or five channels respectively, to the entire code phase. The massive parallel approach [35] extends the EPL structure to cover not only all possible code phases but also the Doppler search space, too.

Therefore, the CCF for all possible code phases and Doppler frequencies are computed individually. Combining the results of these operations lead to the results, exemplarily shown in Fig. 3.4 for an undisturbed scenario and one specific satellite.

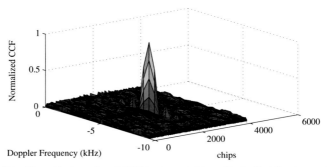

Figure 3.4: Normalized results of CCF implemented in a Single-Shot-Receiver architecture (undisturbed scenario)

The Doppler frequency plays an important role by massively influencing the correlation between the received signal and the local signal replica. This can easily be seen in Fig. 3.4 where the CCF at Doppler frequencies other than the current one does not deliver significant correlation results. This remaining sinusoidal Doppler signal is a residual from the down conversion process in the RF Frontend. The Doppler frequency plays an important role by massively influencing the correlation between the received signal and the local signal replica. This can easily be seen in Figure 3.4 where the CCF at Doppler frequencies other than the current one does not deliver significant correlation results. This remaining sinusoidal Doppler signal is a residual generated by the conversion process from RF to baseband in the RF Frontend.

The frequency difference between actual and nominal carrier frequency originates from the relative movement of the SVs and the user as well as instabilities and noise effects of the Local Oscillator (LO) reference in the RF Frontend [32]. While the former are different for each SV in view, elimination of the individual Doppler frequency in the RF Frontend would only be possible with extensive hardware invest. Instead the received satellite signal is down-converted with the nominal carrier frequency of the GNSS signal and the influence of the Doppler frequency can be eliminated during calculation of the cross correlation function. Second order effects like code-Doppler drift caused by the same effects as carrier-Doppler drift degrading the positioning solution [36] are well examined and can be canceled or at least estimated by common approaches.

Another issue which has a strong influence on the precision of the code phase determination and hardware effort is the channel spacing [37]. In the beginnings of NAVSTAR GPS one-chip spacing was used due to the limited processing power available. Nowadays typical channels are spaced half-chip wise or even smaller.

Once the EPL position and the corresponding results of the CCF are known, an interpolator is applied, which allows prediction of the real CCF maximum. Although there are several possibilities to interpolate the CCF peak [38] [39] at first the easiest, a linear approach, is suitable. Once the real code phases are known, pseudoranges can be calculated with the same algorithms as implemented in the ATR or other common architectures. The equations for determining the actual position from pseudoranges equal those of common architectures [7] [30], which are investigated later in the detailed description of the implementation of the herein proposed architecture.

There are several approaches to compete with the massive hardware usage in the digital baseband domain of assisted architectures. The principle of determining the actual code phase as given in Figure 3.3 can be implemented in two different ways. Despite the straightforward approach of correlating the incoming samples directly with the local replica in time domain, there are more sophisticated implementations of this principle performing the CCF in the frequency domain [40]. This requires the conversion of the incoming signals as well as the local generated replica into the frequency domain, reducing the search for the maximum of the CCF to one multiplication as shown in Figure 3.5. An inverse Fast Fourier Transformation (FFT) of the result of the multiplication delivers the same results as a time domain calculation.

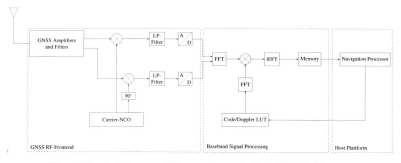

Figure 3.5: Block diagram of Frequency Domain Single-Shot Architecture

From Equation (3.1) which describes the calculation of the CCF in the correlation channel, it can be seen, that correlation by a convolution in time domain corresponds to a multiplication in the frequency domain.

$$CCF_n = \sum_{k=1}^{N} s_{GNSS,k} \; PRN^*_{GNSS,n}$$
$$= \sum_{k=1}^{N} \sqrt{2C} \; d(t)c(t) \; e^{j\phi(t)} \; PRN^*_{GNSS,n} \qquad (3.1)$$

where C represents the carrier power, d the data representing the navigation message, c the ranging code of the SV and ϕ the remaining signal phase of the Doppler residual. $PRN^*_{GNSS,n}$ represents the actual local generated replica of the ranging code of SV n.

3.3   Search Space versus Time-to-First-Fix

The main advantage of assisted architecture like the SSR, its short TTFF, comes at the cost of massive usage of processing power in the digital baseband domain. The hardware effort to be spent is high, irrespective of the implementation of time-domain or frequency-domain architectures. The computation effort concentrates in a huge number of Multiply-Accumulate (MAC) operations which must be implemented, when using time domain approaches. The relaxation performing only one MAC operation in frequency domain implementations is by far compensated by the hardware effort needed for the FFT and Inverse Fast Fourier Transformation (IFFT) operation.

As stated in the former section, a massive parallel approach in the digital baseband processing is implemented in the SSR for calculating all possible code phases of a specific satellite. Therefore the code phase search space is typically searched in half-chip stepping. Meanwhile the search space for the Doppler residuals of ±5 kHz must be monitored, too. Although the Doppler frequency can be approximated based on ephemeris data and approximate position, especially in bad reception conditions, a search with fine Doppler frequency resolution in the remaining Doppler range has to be applied. Remembering NAVSTAR GPS boundary conditions, the code length of 1023 chips corresponds to 2046 channels per Doppler frequency for half-chip wise channel stepping assuming an instant calculation of the actual code phase. Additionally assuming worst-case conditions, the whole 10 kHz Doppler range [13] has to be searched with 500 Hz granularity [41] for a reliable detection of the CCF maximum. The resulting search space spanned by code phase versus Doppler range results in a very high hardware consumption especially since the determination of the actual code phase has to be computed for four satellites. The hardware consumption issue is even intensified when Galileo satellites are considered for the positioning solution because of the longer code length of 4092 chips.

Although the amount of hardware can be cut down by either serializing the search for the code phase of different satellites, serializing the search for Doppler frequencies of a specific satellite or even serializing the search for a specific code phase. All these approaches work at the expense of a longer TTFF, while consuming at least the same amount of power, since the search space is only divided into several sections which are handled sequentially, but not limited in any direction. In order to overcome this problem a new architecture is presented in chapter 5.

CHAPTER 4

Long-Range Time Synchronization

Definition of an official time standard is part of the constitution in various countries around the world. Historically official local time was defined by the local sunrise and sunset but with development of high speed communication the introduction of a common time base was necessary. Since 1972 the official world time represented by UTC is standardized as well as its distribution [42].

UTC and its local predecessors are spread through different wired and wireless channels. While techniques like the transmission of time marks via telephone lines got out of date in modern times, the usage of Network-Time-Protocol (NTP)-timeservers, transporting time information via the Internet, got more and more important. Additionally GNSS receivers, providing a precise local time base tightly coupled to UTC, are used today. Despite the growing importance of internet- and GNSS based time synchronization, the official time information is still broadcasted by radio signals in the LW, Medium-Wave (MW) and Short-Wave (SW) frequency spectrum. There are several European transmitters named after their call sign, like HBG in Switzerland, MSF in England, the French TDF or the German DCF77. Also several stations outside Europe are available like the Russian RWM and RZU or the American WWV, WWVB and WWVH. While the former use Amplitude Modulation (AM) long-wave radio signals in the range between 60 kHz and 152 kHz to spread the time information, the latter use medium and short wave signals up to 20 MHz with the same modulation scheme. This chapter deals with time information transfer and time synchronization via LW radio signals. Especially the unique feature of modulating the phase of the German DCF77 signal and its impact on the accuracy of a synchronization solution is investigated. More information about the SIS of the other time code transmitters can be found in [43], [44] [45] and [46].

4.1   DCF77 Signal-in-Space Characteristics

The DCF77 signal is sent out from a transmitter at Mainflingen near Frankfurt am Main, Germany. It is located at 50°01'56" North, 09°0'39" East using WGS84 coordinates and uses a 150 m high, omnidirectional, top loaded antenna to transmit the signal with 50 kW transmission power. The transmitter emits a 77.5 kHz sine wave which is amplitude modulated in the way that the normalized transmission power is reduced to 25 % at the beginning of each second mark. To transmit binary zeros or ones the amplitude is dropped for either 100 ms or 200 ms. This is illustrated in the upper part of Fig. 4.1. The only exception is the 59th second of each minute when the amplitude is not modulated in order to signalize the beginning of a new minute.

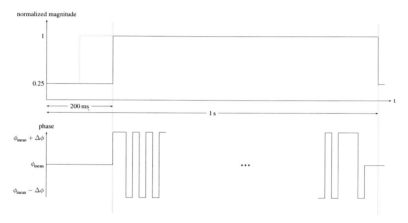

Figure 4.1: Behavior of DCF77 signal in amplitude and phase

In contrast to other time code transmitters the phase of the DCF77 SIS is additionally modulated (see lower part of Figure 4.1). The phase is modulated with a 512 bit long PRN sequence generated from a 9 bit LFSR following equation (4.1).

$$G_{DCF77} = x^9 + x^5 \tag{4.1}$$

The peak phase deviation of the modulation is ±13° starting 200 ms after the second mark, and is clocked with a frequency of 645.83 Hz. Clock frequency and length of the PRN sequence are carefully chosen so that transmission of one epoch is finished within the actual second. The peak phase deviation of ±13°, instead of commonly used 180° for Binary Phase Shift Keying (BPSK), is applied due to the top loaded antenna which otherwise would not be able to follow the modulation. An odd number of chips is used to generate a mean phase of 0° regarding the one second repetition time of the DCF77 signal and therefore avoid disturbances in the usage as a frequency normal [5].

The time information included in the amplitude modulation is again encoded in the phase modulation by either inverting not inverting the PRN sequence. The time information transmitted every minute is illustrated in Table 4.1. The data in bits 20 through 58 is BCD-coded. Furthermore it has to be taken into account that the information actually transmitted is valid for the next minute.

Second	Data
0	Start of minute (always '0')
1 - 14	Weather and disaster control information
15	Callsign bit
16	If '1' : at the end of this hour change between DST and non DST
17	If '1' : DST
18	If '1' : non DST
19	If '1' : leap second is inserted at the end of the actual hour
20	Start of time information (always '1')
21 - 24	Minute (ones)
25 - 27	Minute (tens)
28	Minute parity bit
29 - 32	Hour (ones)
33 - 34	Hour (tens)
35	Hour parity bit
36 - 39	Day of month (ones)
40 - 41	Day of month (tens)
42 - 44	Day of week
45 - 48	Month of year (ones)
49	Month of year (tens)
50 - 53	Year (ones)
54 - 57	Year (tens)
58	Date parity bit
59	no second mark

Table 4.1: Time information encoded in DCF77 SIS [8]

Mathematically the DCF77 SIS can be described as given in equation (4.2), where A_0 represents the RF carrier amplitude, ω the RF carrier frequency and $\Delta\varphi$ the peak phase deviation.

$$DCF77_{RF} = A_0 \left[\sin(\omega t + \Delta\varphi) p_1(t) + \sin(\omega t - \Delta\varphi) p_2(t) \right] \qquad (4.2)$$

The inverse binary sequences $P_1(t)$ and $p_2(t)$ are generated from G_{DCF77} (see equation (4.1)) by equations (4.3).

$$p_1(t) = \frac{1}{2} - G_{DCF77}$$
$$p_2(t) = \frac{1}{2} + G_{DCF77} \qquad (4.3)$$

The resulting spectrum generated from the SIS described in (4.2) can be measured and is given in Figure 4.2 [5]. This will become of special interest, when looking at different receiver architectures the following sections.

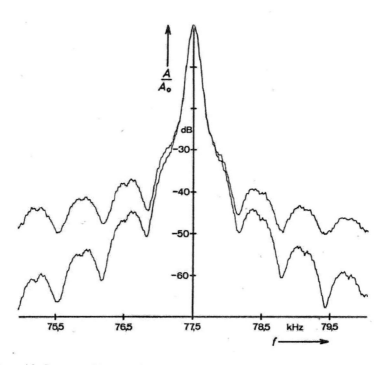

Figure 4.2: Spectrum of the transmitted DCF77 signal (upper: spare antenna, lower: main antenna) [5]

4.2   Time Synchronization via LW Radio Signals

When synchronizing distributed clocks via LW radio transmitters to UTC, the accuracy depends on several parameters such as carrier frequency, modulation scheme and precise knowledge about distance to the transmitter. All mentioned LW time code transmitters use AM to encode the time information, permitting accuracy in the range of several hundred ms. The German DCF77 additionally implements phase modulation as described in detail in the preceding section. Evaluation of the information encoded in the phase modulation of the DCF77 signal allows much higher precision in time synchronization down to the sub-μs range. Generating infrmation with accuracy in that range is essential for the operation of new developed GNSS receiver architecture to be introduced later in this work. Therefore focus is laid on the description of the time synchronization via DCF77 signal in the following sections and chapters.

Despite the fact that radio controlled clocks even synchronized themselves to the DCF77 signal at different locations around the world [47], the DCF77 LW radio signal mainly covers an area of 2000 km around the transmitting beacon at Mainflingen, Germany [8]. The coverage is illustrated in Fig. 4.3. In the range up to 1100 km the ground wave adds the dominant or at least equivalent portion to the received signal strength. At distances further away the LOS path is not available any more and the signal reflected by the iono-spheric d-layer becomes dominant. Although the former mentioned time synchronization reports from around the world rely on several ionospheric and ground scatterings, under common circumstances, time synchronization to the DCF77 signal is only possible when the ground wave or a one time ionospheric scattered wave [8] adds the dominant part to the overall reception power.

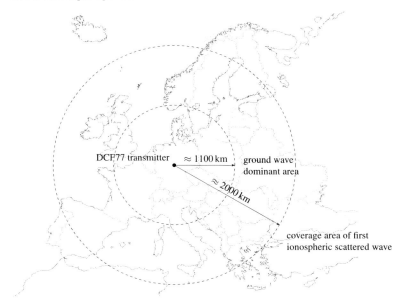

Figure 4.3: Coverage of DCF77 signal (ground wave dominant and one ionospheric scattering)

Each reflection of the DCF77 signal, regardless of the type of scattering (either iono-spheric or ground based), intensifies two issues. First any type of reflection comes along with a significant degradation of signal power. The specific reflection coefficient cannot be easily modeled because it strongly depends on environmental parameters such as time of day, day of year and consequently on properties and condition of the atmosphere, ac-tivity of the sun or humidity and hence conductivity of the reflecting ground [48]. Second every atmospheric scattering introduces an uncertainty in the TOF. Since the signal is reflected at the ionospheric d-layer, which is resided 70-90 km above ground, the specific reflection height is again determined by environmental parameters as mentioned before.

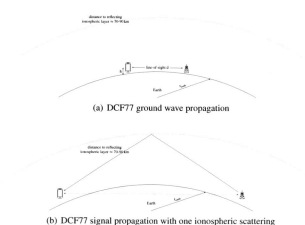

(a) DCF77 ground wave propagation

(b) DCF77 signal propagation with one ionospheric scattering

Figure 4.4: DCF77 signal propagation

Looking at Figures 4.4 it becomes clear that it is essential for a precise time synchronization to eliminate the error introduced by the TOF between transmitter and receiver. Assuming the receiver's position is well known, the TOF error can be canceled. When the ground wave adds the dominant part to the field strength at the antenna, the TOF can be calculated by equation (4.4), where d is the distance between transmitter and receiver and c the speed of light.

$$t_{LOS} = \frac{d}{c} \tag{4.4}$$

Considering the case when ground- and scattered wave field strength are in the same order of magnitude, equation (4.4) is still valid. By examining cases when scattered waves are dominating the field strength at the antenna, the calculation of the TOF error is more complex. This is illustrated in Figure 4.4(b). Because the imaginary LOS does not run along earth's surface, the height above datum h has to be expanded by the distance between the imaginary LOS and the surface of the earth. This leads to a TOF according to Equation (4.5).

$$t_{NLOS} = \frac{2}{c} \sqrt{\left(h + r_{earth} - \sqrt{r_{earth}^2 - \left(\frac{d}{2}\right)^2} \right)^2 + \left(\frac{d}{2}\right)^2} \tag{4.5}$$

The TOF is again determined by the distance between transmitter and receiver d, the height h above datum of the reflection point, the radius of the Earth r_{earth} as illustrated in Figure 4.4(b) and the speed of light c. The TOF for non-LOS situations with multiple scatterings can be calculated by the superposition of several individual scatterings at the atmospheric d-layer, assuming the point of incidence on ground as a new transmitting beacon. The distance d between transmitter and receiver has to be adapted in that case. Residual uncertainty in the TOF results from the inaccuracy in the knowledge about the height above datum of the reflection point which again is determined by the former mentioned environmental parameters [48]. This mainly contributes to the imprecise time synchronization besides the receiver architectural and system immanent errors. Dependent on the receiver architecture implemented the error can be assumed to be better than 100 μs [48].

A common architecture for demodulating AM signals is an envelope detector. To feed the envelope detector the signal has to be received and amplified first. A special issue at LW radio frequencies is the antenna. Due to a wave length in the kilometer range either $\frac{\lambda}{2}$ and $\frac{\lambda}{4}$ dipole antennas or frame antennas are suitable for stationary applications and especially not applicable to be integrated into mobile devices. Instead, ferrite rod antennas are an alternative [49]. As illustrated in Figures 4.5(a) and 4.5(b) the antenna consists of a wire loop which is positioned on the ferrite rod, connected to a parallel capacitor. The ferrite rod and the loop wire form an inductor, which can be tuned by shifting the loop wire across the ferrite rod. Together with the capacitor switched in parallel a LC-resonator tuned to the appropriate resonance frequency is available.

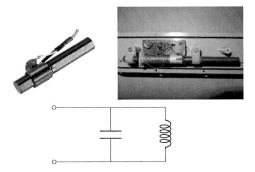

Figure 4.5: Pictures (a)(b) and equivalent circuit diagram (c) of DCF77 ferrite rod antennas [6]

In order to decouple the antenna from the remaining circuit and avoid present heavy load to the antenna which may result in detuning, a high-input-impedance Junction Field Effect Transistor (JFET) is used as the input stage of the receiver. Additionally the JFET stage is used as an amplifier. Depending on the received signal strength several additional amplifier stages may have to be applied. An example block diagram for the reception and demodulation of the DCF77 signal is given in Figure 4.6. The amplifier stages can be realized by discrete transistors as well as operational amplifiers, since LW signals with rather low frequencies have to be amplified. Although the resonant circuit, the antenna is composed of, has a small bandwidth additional filtering has to be applied. Near band interferer whose signal strength is much higher in common reception conditions (e.g. 5th overtone of line repeating frequency of cathode tube displays at 78.15 kHz) can otherwise overdrive the reception circuit.

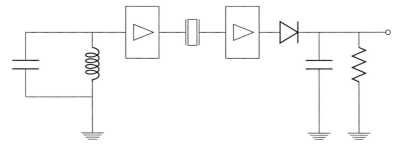

Figure 4.6: Block diagram of common DCF77 receiver Frontend

Remembering the output spectrum of the DCF77 transmitter (see Figure 4.2) only the main lobe of the transmitted signal is essential to recover the information amplitude modulated on the carrier. Therefore a filter with an extreme narrow bandwidth can be applied. For uncritical applications filtering can achieved by a tuning fork crystal, as implemented in Figure 4.6, which has a 3 dB bandwidth at around 10 Hz. Although the AM slope is affected by the narrow filter bandwidth, accuracy is still in the 100 ms range, which is sufficient for most applications. At the output of the detector, realized as a common envelope detector (a diode and a successive low-pass filter), the 100 ms or 200 ms long data marks are provided, which can easily be decoded into the time information by digital signal processing e.g. on a microcontroller.

To achieve an improved synchronization accuracy down to the 100 μs range more complex filter concepts and demodulation architectures have to be applied [50]. Although those architectures have acceptable synchronization errors by only demodulating the AM encoded information, they are not applicable here. Main drawback is the dependency from precise knowledge about TOF which implies precise knowledge about the distance to the the DCF77 transmitter and consequently precise knowledge about the current position of the DCF77 receiver. This information can only be approximated in an AGNSS receiver, since it is the main purpose to generate positioning information. The solution to the problem is to increase accuracy in time synchronization by using the information encoded in the phase modulation of the DCF77 signal. Being an integral part of the later proposed architecture this will be discussed in detail in chapter 5.

CHAPTER 5

DAGNSS Receiver Architecture

After the presentation of common GNSS receiver architectures in chapter 3 and their inherent drawbacks as well as fundamentals of time synchronization via LW radio signals in chapter 4 in the following chapter the functionality of the new developed Double Assisted Global Navigation Satellite System (DAGNSS) receiver architecture is described in detail.

Besides functionality, two key factors drive the development in the consumer market for electronic devices. First the customer's desire for enhanced functionality together with no limitation in usability (especially regarding the endurance of the device running on battery power supply) and second the cut in costs for the manufacturer leading to improved profit margins. Both result in a drive for integration which allows faster and in particular energy efficient products.

The power consumption of mobile communication devices has increasingly has turned more and more into the public focus over the last years. Consequently, the focus not only highlights the complete device but each individual functionality, too. The following sections will introduce a new architecture for a navigation receiver, especially designed to be incorporated in a mobile communication device. Goal of the architecture is to significantly reduce the amount of processing effort required to gain a positioning solution and therefore reduce the power consumption of the device compared to common assisted and non-assisted GNSS receiver architectures.

Recalling the introduction of the ATR in chapter 3, its mean power consumption is mainly determined by the fact that an unlock of the tracking loops in the digital baseband processing has to be avoided. Generally this is achieved by continuous operation of the analog-RF parts as well as the digital baseband signal processing of the receiver. Common AGNSS receiver architectures try to improve the overall power consumption by replacing mean power consumption with high peak power consumption for short operation times. Superiority is gained from very low power consumption during the shutdown interval in-between positioning operations, when substantial parts of the receiver are sent to a sleep mode or even can be isolated from power supply. Several parameters, such as receiver velocity or distance to next turn-point on a navigation route, determine the update rate of the positioning operation in the AGNSS receiver and therefore its wakeup interval. Since no a priori information about the current code phases of the satellite signals is available to the receiver, even in assisted architectures, the whole code phase search space has to be scanned for the actual code phase of each SV. Most of these operations performed to determine the current code phase of the satellite signal are unnecessary, since their results are not used in the positioning solution.

At first a brief system overview will be given. Afterward details of the new architecture, its specific functionality and its operating mode are described in detail. Special focus is laid on the presentation, how to overcome the drawbacks of common architectures, finally proving superiority of the proposed architecture.

5.1   System Overview

The new architecture benefits from the synergistic effect of merging the functionality of two radio receivers. A single-shot based GNSS receiver architecture is augmented by a DCF77 time-code receiver. A block diagram giving an overview of the architecture is given in Figure 5.1.

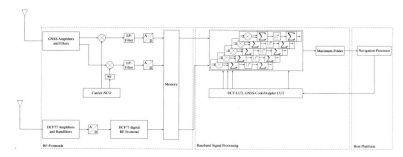

Figure 5.1: Block diagram of newly developed SSR based receiver architecture

The architecture can be divided into four main parts, which will be individually investigated in depth in the following sections:

- GNSS-RF-Frontend

- DCF77-RF-Frontend

- Baseband Signal Processing

- Navigation Processor

Although the genes of the single-shot receiver are still well perceivble, several modifications have to be implemented. While the GNSS RF-Frontend path is left untouched,
compared to other receiver architectures, a second RF-Frontend serving amplification,
filtering and demodulation of the DCF77 SIS is added. To achieve sufficient accuracy
in time synchronization the DCF77 RF-Frontend must be able to encode the information
modulated on the phase of the DCF77 signal.

Besides minor control functions the overwhelming part of the digital baseband processing is the main engine for calculating the current code phases of the satellite signals. This
correlation bank can be adopted with minor modifications from the original SSR architecture. Main difference regarding the digital baseband processing is the introduction of
a FIFO style memory at the input of this block latching the incoming GNSS and DCF77
data for offline processing.

The baseband signal processing engine transfers its results to the navigation processor
which is implemented on the host platform. GNSS assistance data is processed on the
navigation processor, together with the time information gathered via the DCF77 receiver.
Finally the navigation solution is computed.

5.2   Analog-RF Frontend Architectures

5.2.1   GNSS RF Frontend

Similar to the the transition from the ATR to assisted architectures, existing GNSS RF
Frontends can be reused for the DAGNSS receiver, since the new architecture does not
impose special demands to the GNSS-RF Frontend. From that fact only the interface to
the digital baseband processing is of interest, while knowledge about the inner life of the
Frontend is unnecessary.

Recalling explanations in chapter 3, a 2-bit wide digital output is desirable for the analog
frontend. Consequently the GNSS-RF Frontend behaves like a black box, whose Analog
to Digital Converter (ADC) output stage should deliver I- and Q samples, of 2-bit word
width respectively. Additionally the flexibility in the digital baseband processing gives
independence from implementing either zero-Intermediate-Frequency (IF) or a low-IF architectures. Needless to say this is only true as long as the sampling frequency of the
ADCs does not fall below the boundary to fulfill the Nyquist-Shannon-Theorem. Consequently in case of a zero-IF architecture the sampling frequency should be at least
2.046 MHz. In case of low-IF architectures the residual IF-frequency can be removed
together with the Doppler frequency during cross correlation of the GNSS signal.

Based on the research project HIGAPS2, where a zero-IF architecture was chosen, for the
development of the GNSS RF-ASIC (Application Specific Integrated Circuit), without
loss of generality, all investigations here and in the following sections apply to a zero-IF
GNSS-RF-Frontend architectures.

5.2.2   DCF77 RF Frontend

In contrast to the GNSS RF-Frontends its DCF77 counterpart to be used here, has special demands that have to be fulfilled and which are rarely found in commercial solutions. The usage of the information encoded in the AM described in section 4.1 is not suitable here due to the lack of precision in time synchronization. Therefore, the information in the Phase Shift Keying (PSK) modulation has to be evaluated. This disqualifies the already introduced DCF77 receiver architecture (see chapter 4) for the present use case and a new architecture has to be developed. To the knowledge of the author there is only one commercial available product for generating time information using the DCF77 phase modulation [51] which extensively uses analog signal processing.

Due to the carrier frequency of 77.5 kHz a digital approach is more suitable utilizing the conditions available at the input of FPGA (Field-Programmable Gate Array). Still some analog signal processing is necessary for qualifying the signals to be processed in the digital domain. Signal conditioning has to be performed, in particular amplification and filtering. After lifting the signals amplitude to reasonable values filtering is used as a precaution not to break the limit given by the Nyquist-Shannon-Theorem [52] during digitization.

For reception of the DCF77 SIS the type of antenna used is independent of whether the subsequent signal processing is designed to decode the AM or PSK modulated information. Bandwidth required for both information types is sufficient when using a matched ferrite rod antenna. Therefore the type of antenna introduced in chapter 4 can be reused.

Subsequently, the first amplification stage serves two purposes. Besides its main functionality, amplifying the incoming signal, it serves as decoupling stage between antenna and the remaining circuit. An adjacent bandpass filter limits the bandwidth of the received signal not to exceed the maximum allowed input frequency at the ADC. Second another amplification stage is applied, which should be tolerant to high input amplitudes allowing the output signal to clip at the power supply rails without adding distortion especially to the phase of the signal. Now the preprocessed signal can be fed into the ADC for digitization. Again based on the fact that the carrier frequency is as slow as 77.5 kHz the input cell at the pin of an Field-Programmable Gate Array (FPGA) can be used as an 1-bit ADC. Modern FPGAs allow input frequencies of several megahertz, specialized input blocks even higher frequencies in the range of hundreds of megahertz to several gigahertz [53], such that an oversampling of factor 100 and higher of the DCF77 signal is possible. Afterward a straight forward homodyne receiver architecture can be built up in the digital domain. The complete receiver is illustrated in Fig. 5.2.

For the first amplification stage an JFET is predestined due to its decoupling capabilities regarding high impedance sources, which is the case when using a ferrite rod antenna. Circuits using operational amplifiers are conceivable due to the low carrier frequency, too. The adjacent band limitation filter can be realized with simple components, assuming very relaxed requirements to the filter characteristic. An input sampling frequency of 7.5 MHz is chosen for the digital part of the RF Frontend to allow sufficient oversampling to correctly detect the phase deviations of the incoming signal. This corresponds to an attenuation of 80 dB at 3.75 MHz to safely remove any significant contribution of higher frequency portions of the received to the digitized signal. The passband corner frequency of the band limitation filter is assumed to be 1 MHz. Since there are little services using frequency bands in the range the DCF77 carrier frequency, together with the inherent filter functionality of the ferrite rod antenna and the homodyne receiver concept in the digital domain, a low pass filter characteristic is sufficient here.

After digitization the received DCF77 signal is mixed with a 77.5 kHz reference signal generated by a Direct Digital Synthesis (DDS), removing carrier frequency and demodulating the information encoded in the phase of the DCF77 signal. The subsequent filter stage, removing unwanted mixing products of the downconversion and demodulation process, additionally makes the mixing process insensitive to the type of the waveform of the reference signal. The mixing products between the overtones of the LO and the incoming DCF77 signal result in spectrum components which are located at much higher frequencies than the effective bandwidth of the demodulated DCF77 information. Using an In-Phase Quadrature (IQ) mixer architecture the dependency of the demodulation process from the absolute phase difference between RF and LO mixer input can be removed in the digital baseband processing after cross correlation.

The filter stage not only removes unwanted spectrum components generated during the demodulation process, but can be used for downsampling, too. While processing of the demodulation must be performed at a suitable sampling frequency (again according to the Nyquist-Shannon-Theorem at least twice the RF signal frequency) the data rate of the information is much slower, allowing much lower sampling frequencies in the adjacent baseband signal processing. Of course, the filter characteristic has to be adapted to the rate of the demodulated data, too.

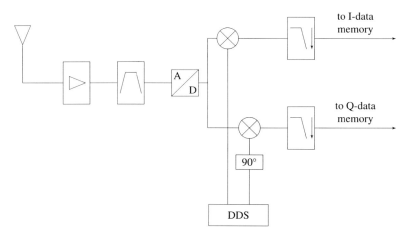

Figure 5.2: Block diagram of DCF77 RF Frontend for PSK demodulation

5.3   Digital Baseband Signal Processing Architecture

The main improvement is achieved in the digital baseband signal processing when comparing common assisted and the introduced architectures in this work. Therefore, core functionality of the SSR, namely applying a matched filter approach to the satellite signals is adapted. The matched-filter is realized by computing the cross correlation function between the incoming signals and their local generated replica. While satellite signals are exclusively fed to the cross correlation bank in the SSR architecture, additionally there is the possibility to reuse the block for calculating the current code phase of the DCF77 signal.

As stated in section 3.3 there is a strong dependency between the size of the code phase search space and the TTFF. Reduction in TTFF at a given search space size can be achieved by either increasing parallelization of the cross correlation engine or by reducing the search space. Extensive parallelization impedes the goal of an energy efficient receiver architecture, which means that precautions for limiting the code phase search space must be taken.

The assistance information gathered by the host's cellular connection allows the limitation of the search space to specific satellites and an approximation of the actual Doppler frequency for each SV in view. Nevertheless, the determination of the exact code phase demands the computation of the cross correlation function for every possible code phase using a given spacing. Estimation of the amount of channels which have to be applied for a full parallel architecture can be seen in Table 5.1. Determination of the exact code phase of any GNSS satellite is consequently very expensive in terms of hardware consumption. This issue is intensified especially regarding the use of 2nd generation GNSS signals with longer code lengths.

System	Channel spacing	Doppler steps	No. of channels
GPS	1	20	20460
GPS	$\frac{1}{2}$	20	40920
Galileo	1	20	81840
Galileo	$\frac{1}{2}$	20	163680
DCF77	1	1	646
DCF77	$\frac{1}{2}$	1	1292

Table 5.1: Estimation of the number of CCF channels for determining exact code phase signals with different channel spacings and Doppler frequencies

All calculations for the amount of channels needed to determine the current maximum of the cross correlation function are based on the assumption that all possible phase shifts have to be searched.

Instead of that, knowledge about time gives the opportunity to shrink the search space in the second dimension, the code phase. Therefore some modifications compared to common GNSS receiver baseband signal processing architectures are unavoidable.

The realization of computing the cross correlation function between the incoming signals and an appropriate local generated replica (either DCF77 or GNSS PRNs) in the present architecture consequently has four core blocks:

- Input Signal memory

- PRN memory

- Control

- Correlation engine

5.3.1   Input Signal Memory

Several steps of the algorithm to determine to actual position cannot be parallelized, because they must be executed sequentially. In particular time information is generated first. Afterward satellite signals are processed. This requires offline processing of the GNSS signals. Therefore, it is inevitably to buffer the incoming signal samples phase aligned in a memory prior to processing. Both, the DCF77 and the GNSS signal, have to be recorded for an appropriate period, at least one PRN code epoch each.

It is absolutely essential to start sampling and storing both signals in parallel. Only parallel reception of the incoming satellite and DCF77 signals and their consolidation by a single buffering start point allows the precalculation of the satellite signal and subsequent offline processing. Depending on input word width and integration time, memory usage for buffering the incoming signals can be very high. The amount of necessary memory bits N, can be estimated by equation (5.1). The input word width is defined by n_{iw} and s_x represents the number of samples to be recorded for the satellite- and the DCF77 signal, respectively. Both s_{GNSS} and s_{DCF77} are composed of the epoch lengths, 1023 samples for GPS, 4092 samples for Galileo and 646 samples for DCF77 and the appropriate oversampling factor.

$$N = 2 \left(n_{iw} s_{GNSS} + n_{iw} s_{DCF77} \right) \tag{5.1}$$

The amount of memory which must be provided for storing the incoming data signals is negligible in modern digital signal processing architectures, regardless of the implementation either in a FPGA or an Application Specific Integrated Circuit (ASIC). Inserting reasonable values into equation (5.1), e.g. n_{iw} = 2, s_{GNSS} = 16386 samples and s_{DCF77} = 1292 samples, results in a memory block for storing the incoming signals of W_{mem} of approximately 4.4 kByte.

The memory block, buffering the incoming DCF77 and GNSS signals, can be arranged in a true dual port memory structure with several pages. Characteristic for dual port memory architectures are different input and output clock domains which comes along with several advantages. The organization in several memory pages allows the separation of navigation and time-code signals using different input sampling clock frequencies. Additionally the offline processing of the stored data in any page can be performed with a single, much higher system clock frequency. The optimal organization of the input memory pages would be in a $(n_{iw})x(s_x)$ layout, where x stands for either GNSS or DCF77, as depicted together with inputs and outputs as well as the appropriate multiplexers in Figure 5.3.

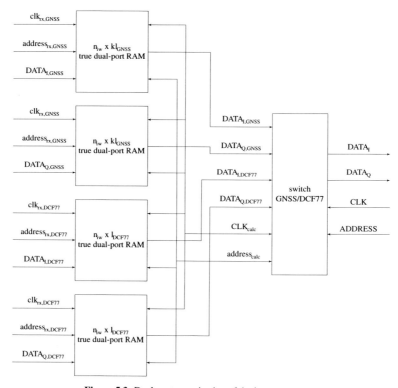

Figure 5.3: Dual-port organization of the input memory

Comparing the DCF77 and GNSS memory pages two differences stand out. Besides the initial length of memory organization either l_{GNSS} or l_{DCF77} another parameter N is introduced to determine the actual length of the GNSS memory page. This parameter defines the number of epochs the memory page is able to collect. The parameter has to be applied only to the GNSS memory pages since determination of the cross correlation function for the DCF77 signal in any case must only be performed over one signal epoch. Regarding carrier frequency and reasonable user positions, it is hard to imagine reception situations where the cross correlation function of the GNSS signal (even when a massive amount of signal epochs is used) does deliver significant contributions, while the determination of the CCF over one epoch of the DCF77 signal does not. Therefore it is not reasonable to spend additional hardware, by expanding the DCF77 memory pages.

Another possibility for implementing the memory pages are First-In-First-Out (FIFO) style memory blocks. Compared to the implementation using the dual-port memory architecture a FIFO based implementation as shown in Figure 5.4 saves hardware effort by cutting the read- and write address generation. Instead the hardware costs accumulate in the input data and clock multiplexers and the feedback lines.

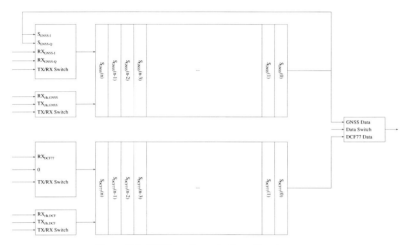

Figure 5.4: FIFO based organization of input memory

The control part of the baseband signal processing supervises the input memory and stops recording of baseband samples after a predefined reception time or after recording the corresponding number of input samples.

5.3.2   Local PRN Memory

The organization of the memory storing the samples of the local generated replica of the signal to be correlated distinctly follows other rules than the input memory. While only one sample of the incoming signal is fed to all cross correlating channels at a particular time, the content of the local PRN replica memory must be spread over all cross correlation channels such that every adjacent channel is provided with the adjacent replica sample. A Random-Access-Memory (RAM) only allows access to one (or in case of dual-ported RAM architectures two) word(s) stored in the memory at a time. Since the cross correlation shall be performed for all channels in parallel, which requires full parallel access to all replica samples at every particular time, this kind of architecture is not convenient. Instead a FIFO approach with full parallel out is more reasonable.

A block diagram illustrating a possible realization of the memory accommodating the requirements for the local PRN replica is given in Fig. 5.5.

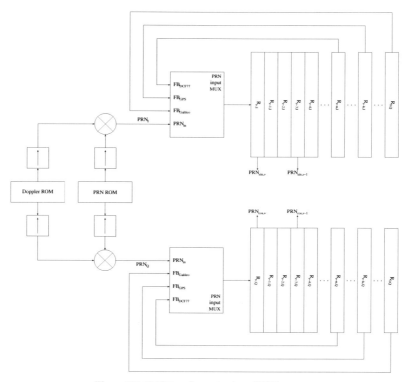

Figure 5.5: FIFO based organization of PRN memory

Original GNSS, DCF77 PRN sequences and initial Doppler data are stored in Read-On-ly-Memory (ROM) blocks. Two possible paths of operation depending on the assigned command can be followed. When the memory block is assigned to fill the FIFO memory, the appropriate PRN sequence and Doppler data is read sample per sample from the ROM blocks. The information is fed to a mixer providing the possibility not only to generate Doppler compensated GNSS or DCF77 PRN replicas but also to upsample the resulting sequences to the sample rate of the received signals.

The generated sequence is fed to the FIFO memory until the end of the current epoch is processed completely. This approach together with the random read access capability of the PRN ROM allows easy application of the precalculated code phase offset to the replica PRN generation. This is realized through an additional address offset while reading the PRN data.

Since all three possible signal types (NAVSTAR GPS, Galileo and DCF77) differ in se-quence length the input multiplexer must not only be able to switch between incoming data from the mixer during the fill operation, but to distinguish between the several feed-back inputs. The feedback of data samples expands the functionality of the FIFO memory architecture to a feedback shift register while the baseband signal processing is calculat-ing the CCF. Two advantages arise from the given architecture. A single feedback shift register is sufficient to feed all cross correlation channels in the correlation bank. First, by inserting output taps at the appropriate positions of the register, namely at positions over-sampling factor times channel spacing, all correlation channels can be fed with data in parallel. Second the replica data can be used for the correlating several epochs of GNSS input data with only one epoch of replica data, consequently massively decreasing the amount of memory which must be available.

Not all words stored in the PRN replica memory must be provided to the correlation bank. Instead only the first part of the PRN-FIFO memory is exported in parallel to the cross correlation bank. The number of words to be provided to the cross correlation bank is determined by the length of the DCF77 replica, which comprises 646 samples and the spacing between the cross correlating channels in terms of samples regarding the original PRN replica. The number of taps directly corresponds to the number of channels in the cross correlation bank. For correctly connecting the taps to the FIFO memory structure the sampling frequency of the incoming signal has to be taken into account, too. For sake of simplicity it is assumed that the sampling factors calculated from the division of the sampling frequency of either GNSS or DCF77 signals by their appropriate modulation frequency are congruent. This limitation has to be accepted since the hardware effort to realize the switching matrix, allowing the FIFO taps output data to be multiplexed to the appropriate cross correlation channel would be unmanageable high. The same argumen-tation must be made when thinking about adaptive sampling frequencies, especially for the GNSS signals adapting the current reception conditions, although the RF Frontend architectures as well as the cross correlation bank does not prohibit such approaches. The implementation done in this work emanate from a channel spacing of 0.5 chips and a sam-pling factor of four. This results in 1292 correlation channels and therefore taps provided linearly distributed over the first 5168 words the FIFO memory is built up from.

Replacing the length of the DCF77 PRN replica with the maximum replica length in this calculation, hence 4092 samples in case of searching the maximum of the cross correlation function for a Galileo signal, the length of the complete FIFO memory can be calculated, too. The result corresponds to the amount of hardware effort to be implemented in a com-mon SSR architecture to calculate the result of the CCF. Applying the already mentioned values for channel spacing and sampling factor the calculated memory width, and hence the the number of cross correlation channels, comprises 32736 words.

Although it was stated in in subsection 5.2.1 a zero-IF architecture is assumed for the GNSS RF Frontend, the flexibility of the architecture, here proposed for the first time, al-lows the adaption to low-IF architectures, too. Besides modifying the size of the memory storing the input samples due to an adapted sampling rate of the input signals, still fulfill-ing the Nyqist-Shannon theorem, the only difference to be applied to the baseband signal processing is due to the generation of the local PRN replica. In particular the change of the boundary condition applies to the content of the Doppler ROM. While in case of zero-IF architectures the content corresponds to the sine wave samples with frequencies up to ±5 kHz, in case of a low-IF architecture its content must be modified to reflect sine waves representing frequencies of $f_{IF} \pm 5$ kHz. The stepping between the Doppler frequencies must not be touched, such that no modification to the size of the Doppler ROM must be made. Of course the positions of the taps added the FIFO memory must be adapted, which can easily be parametrized in VHDL during the synthesis process of the VHDL compiler.

5.3.3   Correlation Engine

The correlator engine is built up from several correlator channels, each calculating the cross correlation function between the incoming DCF77 or GNSS signal and a specific code phase of the generated corresponding replica. Each correlator channel is composed of a coherent and a non-coherent integration part. Even more sophisticated correlator architectures have been developed which e.g. implement additional differential correlation techniques [54]. These architectures give superior possibilities to determine the maximum of the CCF under bad reception conditions, hence allowing the determination of a positioning solution under these conditions, too. However the increased sensitivity does not justify their increased complexity. Therefore the correlation channel initially proposed by the HIGAPS project [55] which is shown in Figure 5.6 provides sufficient sensitivity to be used as the basis for the SSR as well as for the correlation bank of the here proposed architecture.

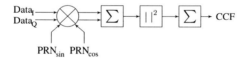

Figure 5.6: Correlator channel implementing coherent and non-coherent integration

The correlator channels in the baseband signal processing of the SSR are arranged in parallel to form the correlator bank as shown in Figure 5.7, as they are in the present architecture. As partially already introduced in the previous section 5.3.2, the maximum number of channels is dependent on the length of the DCF77 PRN replica as well as the channel spacing. Since no a priori information about the code phase of the DCF77 signal with respect of its replica is available, still the whole code phase for that signal has to be searched. Assuming a channel spacing of 0.5 chips, the maximum number of channels ν needed to be provided is

$$\nu = 646 \, \text{chips} \, \frac{1}{0.5 \, \text{chips/channel}} = 1292 \, \text{channels}. \tag{5.2}$$

Essential difference in the correlation bank compared to common architectures, is the adaption of possible input signals to the coherent integration at the RF as well as LO paths. Now, not only GNSS signals are legal assignment to the inputs but the DCF77 signal, too.

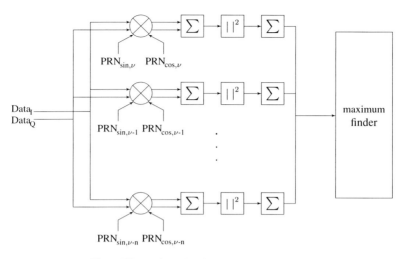

Figure 5.7: Baseband signal processing correlator bank

While the use of coherent integration in the correlation channel is straight-forward, reasons for embedding a non-coherent integration are not visible at first glance. As introduced in chapter 2 the navigation message is encoded in the GNSS SIS by transferring either inverted or non-inverted PRNsequences. As proved later, it is possible to precalculate an approximate value of the current code phase but the value of the corresponding data bit cannot be predicted. Therefore an adaption of the local generated replica to the data bit value is not possible. It has to be taken into account that the CCF delivers the same magnitude result but with inverted sign, applying either inverted or non-inverted replicas. Consequently performing the coherent integration over data bit boundaries together with a change of the data bit value can lead to a degradation of the result of the CCF. Because of length of the NAVSTAR GPS data bit of 20 ms the maximum integration time would be limited to that value. Additionally this assumption would only be valid having perfect synchronization to the navigation message data bit. However, longer integration times may be necessary in bad signal conditions.

To overcome this problem it is inevitable to buffer the result of the coherent integration and transfer it to the noncoherent integration. While the results for the coherent integration are determined separately for I-/Q samples, prior to the noncoherent integration the absolute value of the resulting vector is calculated. This mainly serves the purpose of removing a possible phase difference between GNSS SIS and the Doppler frequency at the LO signal of the coherent mixer.

Together with equation (4.2) the cross correlation of N input samples of the DCF77 signal follows equation (5.3). The result of the j-th correlation channel CCF_j is therefore given by the multiplication of the received input samples DCF_{input}, equal to the description in equation (5.3), and the j-th bit of the local generated replica register containing DCF_{pm}, initialized by the function given in equation (4.1). The result of the multiplication is summed over N samples, while the received DCF77 signal samples are shifted through the input FIFO. N must be equal to at least one complete code phase to assure acquisition of the DCF77 code phase. The number of channels j is defined by the number of chips per code cycle n_{cc} (DCF77: n_{cc} = 646 bits) and the channel spacing k (often k = $\Delta \{PRN_j, PRN_{j-1}\}$ = 0.5 chips).

$$CCF_j = \sum_{k=1}^{N} s_{DCF77,k} PRN^*_{DCF77,j}$$

$$j = [1 \ \frac{n_{cc}}{k}] = [1 \ \frac{646}{k}]$$

$$(5.3)$$

The spacing between the channels results in a time difference between two adjacent channels of according to Equation (5.4).

$$\Delta t_{j,j-1} = \frac{793 \, \mu s}{\max \{j\}} \tag{5.4}$$

Again j is the number of cross correlation channels as defined in equation (5.3). Under ideal circumstances the operation described by equation (5.3) delivers the exact offset between the local RTC and UTC. It is implied that sampling of the DCF77 signal is done coherently with the steps of the phase modulation chips, which cannot be guaranteed, moreover most cases do not cover this situation. Consequently, the global maximum of the correlation function must not be congruent with the maximum calculated by equation (5.3). This circumstance and its solution is further investigated in section 5.5. Since this problem is solved in software on the host, transfer of the CCF results must be realized. As it will be investigated later, depending on the type of interpolation function to be applied, different data types and datarates for the communication channel between digital baseband processing and host hardware have to be realized. To ease the demands of the transfer protocol as well as the amount of data to be transferred, prior to the transfer, a maximum finder circuit is applied to the results of the CCF. In case of an interpolation engine using a linear approach only the channel number calculating max{CCF_j} as well as the preceding and successive channel numbers are relevant. Additionally the corresponding CCF values are of interest (remembering the EPL structure as described in section 3.2). Consequently, the transfer operation can be limited to the given six values. For the search of the CCF maximum a serial approach according to Figure 5.8 is pursued.

Initial channel number and the corresponding result of the cross correlation function for channel j = 1 are loaded to the present channel and present value registers, respectively, as shown in Figure 5.8. Then CCF results of channels j = [2 n_{cc}k] are compared iteratively to the value stored in the present value register. In case the content of the present value register is smaller than CCF value in the channel selected for comparison, both, the contents of the present channel and the present value register are replaced with the values from the currently selected channel. After running through all channels the present channel and present value registers must store the channel which calculated the maximum CCF and the corresponding channel number. Early and late channels and their appropri-

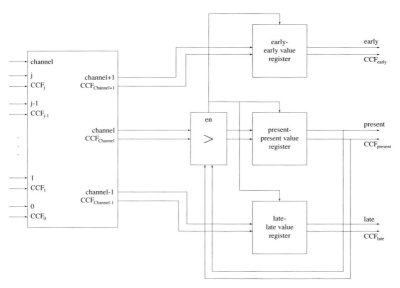

Figure 5.8: Maximum finder circuit

ate values are derived in parallel during the iterative search for the present channel. The early/early value and late/late value registers are filled with the preceding and successive channels of the current present register and their corresponding CCF results. This operation mode is only applicable in case that a linear interpolation approach delivers sufficient precision for determination of the real maximum of the CCF. In case of applying for example a sinc-interpolation technique the whole result vector must be transferred to the interpolation block.

The approach for determining the coherent cross correlation function $cCCF_{x,y}$, using the architecture illustrated in Figure 5.7, can be rewritten using equations 5.3 to (5.5).

$$cCCF_{x,y} = \sum_{n=1}^{N} s_{x,n} PRN_{x,res}$$

$$y = [1, n_{cc}512]$$

$$N = \begin{cases} k\,512, & \text{when x = DCF77} \\ k\,102320 & \text{when x = NAVSTAR GPS} \\ k\,4092, & \text{when x = Galileo} \end{cases}$$

(5.5)

The last step in order to let equations (5.5) correspond to Figure 5.7 is adding non-coherent integration over p coherent integration cycles each correlating N samples for I- and Q-channels. This leads to equation (5.6). While in theory p can run from one to infinity of course infinity is only a theoretical value. If the non-coherent integration does not deliver significant results after reasonable time, the signal of the appropriate SV is obviously not present in the actual received signal.

$$ncCCF_{x,y} = \sum_{p=1}^{M} \sqrt{ \left[\sum_{n=1}^{N} s_{x,n,I} PRN_{x,res} \right]^2 + \left[\sum_{n=1}^{N} s_{x,n,Q} PRN_{x,res} \right]^2 } \qquad (5.6)$$
$$M = [1, \infty]$$

Inserting equation (3.1) and (5.5) into equation (5.6) it is obvious that assuming a perfect approximation of the Doppler drift, the influence of the Doppler residual is eliminated from the CCF by the Pythagorean theorem.

5.3.4   Baseband Control

Main tasks of the Baseband Control block are:

- Implementation of the communication between digital baseband processing and host platform.

- Generation of command and control signals for the digital baseband precessing.

Based on the information transferred, the Baseband Control block fills the memory for the local PRN replica with the result of the multiplication between appropriate Doppler frequency - and PRN sequence. Additionally the control block supervises the calculation of the CCF and the reception of both, DCF77 and GNSS signals.

Triggered by an interrupt signal the reception is started in parallel and stopped after the desired individual amount of samples for each signal are stored in their dedicated memory. While calculating the CCF the address generation for both the PRN replica memory and the memory storing the received samples are handled by the Baseband Control block.

During calculation of the CCF the Baseband Control sets the registers for the transfer of intermediate results of the coherent and non-coherent integration. Additionally, once calculation of the CCF results is finished the transfer of these results to the host platform and the signalization of the end of the calculation operation are supervised by the Baseband Control block.

5.4   Host Hardware and Operating System

Primary assignment of the host hardware is to provide the basis for the operating system, hence the positioning software and the appropriate communication interfaces to the cellular network, to the baseband signal processing and Human Interface Devices (HIDs).

Although there exist numerous possibilities to build up the host, the mobile nature of the receiver architecture, which is reflected by the goal of minimizing the power consumption by enhancements in architecture, calls for the usage of common embedded hardware components and operating system.

Since there are a lot of possible embedded hardware solutions available, each with different requirements and peculiarities, it is almost impossible to describe functionality and specifics in detail. Nevertheless, some general statements about the demands to the hardware can be made. The embedded Central Processing Unit (CPU) must be able to perform the complex mathematical operations executed by the positioning software and should be able to calculate those functions in reasonable time. At the same time, not only processing the power, but also sufficient memory for the processor must be available, to buffer essential variables during the positioning operation. These variables include intermediate data while calculating the position as well as semi-permanent data like the navigation message. Therefore, very simple and hence low-power microcontrollers are disqualified. Instead mobile CPU derivatives based on for example ARM [56], Intel Atom [57] or similar architectures with more complex instruction sets should be used.

The implementation of general purpose processors allows the use of an operating system since software does not run directly on the hardware (in comparison to microcontrollers). The operating system abstracts access to the hardware components and allows the developer to write the software performing the positioning operation in a high-level programming language. This again gives additional degrees of freedom to the developer since the software can make use of predefined libraries especially for controlling interfaces, memorys or attached devices. There are numerous possibilities for embedded operating systems, but the one used for this particular architecture is Google Android [58], which can be ported to almost all standard embedded systems. Besides the portability to different platforms Android comes along with compilers for C/C ++ which allow a platform independent development of the positioning software and validation on the given host platform. Last but not least Google Android is an industry standard embedded operating system, with focus on application in mobile architectures.

5.5   Positioning Software

The positioning software is responsible for the calculation of the positioning solution using distances to four SVs. The pseudo-range, hence the distance to a specific satellite, is derived after applying several compensation algorithms to the result of the corresponding cross correlation function. The output of the appropriate CCF is the code phase for the specific SV, calculated by the baseband signal processing according to equation (5.6).

Prior to starting the CCF operation, as stated in section 5.3, the baseband signal processing must be initialized. This initialization information comprises the PRN number as well as the initial shift for cross correlation of received GNSS signals. This information is gathered based on navigation message, time- and position assistance data. Hence, the retrieval and parsing of the assistance data is included in the tasks of the positioning software.

The flow chart in Figure 5.9 illustrates the algorithm which the software is performing during the positioning operation.

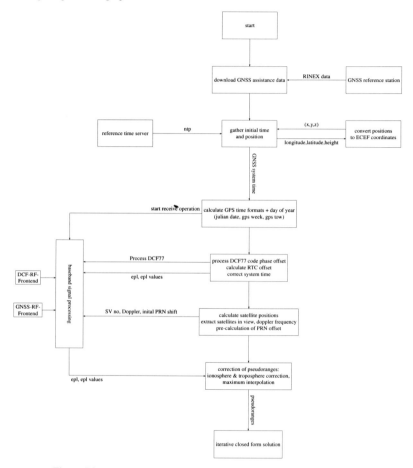

Figure 5.9: Flowchart of software algorithm to determine the actual position

The algorithm starts with a rough estimation of time and position data. As introduced earlier this information is gathered from the assistance cellular network. Some boundary conditions have to be paid attention to while gathering the initial time and position approximation. The accuracy of the initial system time estimation has to be better than 1 second to solve the ambiguity due to the 1 second periodicity of the DCF77 PRN sequence. The 1 ms period of the NAVSTAR GPS PRN sequences again introduces ambiguity, now in the positioning solution with a periodicity of around 299.792 km. Hence the error in initial position estimation should not exceed that radius. The initial position approximation serves another purpose, too. As specified in chapter 4 the TOF of the DCF77 signal has to be taken into account, when evaluating the time information. Together with the initial position information a sufficient TOF estimation can be provided, adjusting and defining the time information of the DCF77 signal.

The initial position information in this architecture is derived from the cellular network connection. The position of the base station, the device is logged onto, is used as the initial position approximation. More sophisticated ways of determining a position approximation utilizing the possibilities of cellular networks include triangulation with several base stations, Time Difference of Arrival (TDOA) approaches on the cellular network signals or the analysis of Received Signal Strength Indicator (RSSI) data generated by the cellular Frontend [59]. These possibilities give the opportunity to even more precise approximations of the initial position and a better determination of the TOF of the DCF77 signal. However, it will be demonstrated in the following that approximating the user's position with the position of the cellular base station is sufficient for this architecture.

Considering a mobile device logged onto a Global System for Mobile Communications (GSM) base station on the countryside, supplying a region of 20 km in diameter with cellular network connection, which is approximately the biggest cell size used in GSM mobile communications networks (and at the same time the biggest cell size used in any cellular network available today). The resulting uncertainty in time synchronization to the DCF77 clock is then mainly determined by two factors. First time error immanent to the system as described in chapter 4 of maximum 250 ns and the error introduced by the imprecision in the TOF approximation of the DCF77 signal. The maximum of this second error t_{error} can be approximated together with the speed of light c, for the assumed cell size of 20 km in diameter according to equation (5.7).

$$t_{error} = \frac{r}{c} = \frac{10000\,\mathrm{m}}{299792458\,\frac{\mathrm{m}}{\mathrm{s}}} \approx 33,35\,\mu s \qquad (5.7)$$

Assuming ideal conditions, hence precise time information and SVs positions, the number or cross correlation channels required to find the maximum of the CCF would be reduced to one. The effect of the remaining DCF77 TOF error as well as imprecise SV positions result in a deteriorated prediction of the expected SV code phases and therefore a widening of the search window for the exact code phase of the SVs. Although the TOF error has to be taken into consideration when determining the actual code phase of the satellite signals, it does not deadly affect the architecture. Assuming the worst case scenario of 33.35 µs error in the TOF measurement of the DCF77 signal, this obviously corresponds to an inherent imprecision in the precalculation of the SVs code phases of 33.35 µs or approximately 34.15 chips. But the imprecise knowledge about the TOF of the DCF77 signal adds another issue to the precalculation of the GNSS code phases, too. Since the satellite is moving constantly the uncertainty in time corresponds to an uncertainty in position of the satellite. As depicted in Figs. 5.10(a) and 5.10(b) the error in the position approximation for the SVs becomes maximal for low elevation angles.

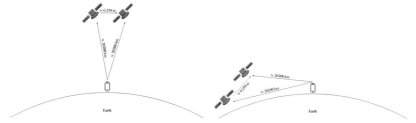

(a) Code phase error, satellite with high elevation angle (b) Code phase error, satellite with low elevation angle

Figure 5.10: Code phase error approximation for satellites at high and low elevation angles

$$x_{error} = v_{satellite} \, t_{err,max} \approx 0.389 \, \frac{\text{m}}{\text{μs}} \, 66.8 \, \text{μs} \approx 26 \, \text{m} \tag{5.8}$$

$$n_{error} = \left[d_{u-sat} - (d_{u-sat} - x_{error}) \right] \frac{l_{PRN}}{c \, t_{PRN}} =$$
$$= 26 \, \text{m} \, \frac{1023}{1 \, \text{ms}} \, \frac{1}{299 \, 792.458 \, \frac{\text{m}}{\text{ms}}} = 0.09 \, \text{chips} \tag{5.9}$$

The GPS satellites, moving at approximately $3.89 \, \frac{\text{km}}{\text{s}}$ [60], are covering a distance during this time period of 0.259 m according to equation (5.8). For low elevation angles the distance traveled by the satellite directly corresponds to the error in position calculation for the transmission position of the SV. Applying the position error to equation (5.9), the conversion of the distance to a resulting code phase error leads to 0.09 chips.

A third source of inaccuracy is introduced after calculating the cross correlation function regarding the DCF77 signal. Due to the finite sampling rate an interpolation algorithm has to be applied to the output of the CCF, since synchronization between input signal and sampling frequency at the input of the digital baseband processing cannot be guaranteed. Even when no interpolation is applied, which reflects the worst case error approximation, the inaccuracy cannot exceed the channel spacing (hence $\frac{1}{2}$ chip in the present architecture).

Adding worst case errors for GNSS code phase approximation for all three issues leads to a maximum error of around 70 chips, which is still well below the number of channels that are provided in the current architecture, due to the 646 chips code length of the DCF77 PRN sequence.

As stated earlier, the architecture gains its superiority over common architectures by synchronizing the clock of the mobile device to a reference station prior to the positioning operation. Essential to the first coarse synchronization operation, performed via cellular network connection, is to overcome the ambiguity inherent to the DCF77 signal. Of course, if the confidence to the information of the built-in RTC is sufficiently high, this information can be used, too. E.g., the cellular connection attached to the mobile device, allows the gathering of the initial time information the cellular interface. Several solutions realizing the coarse synchronization between system time and an appropriate reference is to synchronize the system time with a remote time server via the NTP. Remote time servers are locked to a very stable time reference such as an atomic clock delivering highly precise time information. The NTP [61] is a protocol defined to transfer time information between computer systems via a TCP/IP connection regardless of the underlying layers of the OSI-model [62]. It allows the synchronization of two clocks down to an accuracy of around 10-20 ms in common networks (either LAN, WLAN, UMTS). This is sufficient since the initial time information is only needed to resolve the 1 second ambiguity in the time information decoded from the DCF77 signal.

Subsequently, the main software algorithm starts with a conversion of the coordinates of the initial position estimation. Since these coordinates (the position of the base station) are commonly given in longitude/latitude/height coordinates, defined in WGS-84, these have to be converted to the ECEF coordinate system.

Meanwhile the software changes the state of the baseband signal processing unit into reception state. Parallel to the receive command for the baseband signal processing, when data recording is started, a time stamp has to be generated based on the local RTC. Reception of DCF77 and GNSS signal samples stops when a certain amount of data is collected. The period to store data must comprise at least one epoch of each kind of data, DCF77 and GNSS. Hence, a number of DCF77 samples have to be stored to an equivalent of 1 second, while the number of samples for NAVSTAR GPS and Galileo must exceed the equivalent of 1 ms and 4 ms respectively.

Once the received data is available, the software advises the baseband signal processor to perform the cross correlation function on the stored DCF77 signal samples and its corresponding local generated replica. Depending on the time stamp generated, based on the information gathered via cellular network, as well as forcing the calculated data to reflect only the offset between local system time and UTC, two solutions are imaginable. Either signal reception is started at the local second mark, directly delivering the desired code phase offset. This approach extends the TTFF by the time until reception can be started at the next second mark. Second possibility is to gather the information about the approximate delay of local system time to UTC by transferring the initial time approximation to the baseband signal processor along with the command to calculate the CCF. The second solution would additionally allow a precalculation of the code phase of the DCF77 signal, resulting in reduced effort for calculating the CCF. However, the shift distance of the maximum CCF value is corresponding to the offset of the local RTC with respect to UTC.

Due to the missing synchronization of the sampling and received signal carrier frequency, the real maximum of the CCF does not necessarily match the calculated maximum. Therefore an interpolation algorithm for the maximum of the CCF has to be applied.

Besides the de facto realization the prior decision to be made is whether this algorithm must be implemented in hardware or software. Although this decision can hardly be decoupled from the real algorithm some severe arguments are determining the decision. Mainly the communication interface between baseband signal processing and the host processor and especially the available data rate is one of the key factors. If the interpolation algorithm is realized in software, consequently all output values of the cross correlation channels have to be transferred to the host processor. Depending on the type of communication link, this can not only consume a significant amount of time but the communication must not be absolutely fail-safe. Additionally considerations regarding speed would make it desirable to calculate the algorithm in hardware. But since a hardware implementation consumes significant resources, a software implementation seems more reasonable.

The architecture at hand implements data transfer between host platform and baseband signal processing through a simple but fast Serial Peripheral Interface (SPI). Nevertheless the serial nature of the interface denies a software implementation, since not only data has to be transferred serially but the word width representing the absolute value of the results has to be split, in order to fit the communication link word width. This again increases the number of transfers to complete a communication cycle. Therefore two possible implementations are investigated in this work. On the one hand a linear approximation algorithm is implemented, realized in hardware. Second a sinc-interpolation algorithm is realized in software.

The most simple solution for the interpolation algorithm would be the linear approach, being implemented as illustrated in Figs. 5.12. Again it has to be mentioned that the time offset can be calculated according to equation (5.10), converting the shift distance of the DCF77 replica between zero shift (representing local system time) and the present channel to a time information. Due to the discrete channel spacing in the baseband signal processing and the non-coherent sampling of the input signal, the real maximum of the CCF must not necessarily correspond to the calculated maximum as illustrated in Figures 5.11 assuming worst case errors.

$$t_{error} = 2t_{cl,DCF77} \left(\frac{n_{cc,DCF77}}{k} \right)^{-1} = 2793\,\text{ms} \left(\frac{512}{k} \right)^{-1} \tag{5.10}$$

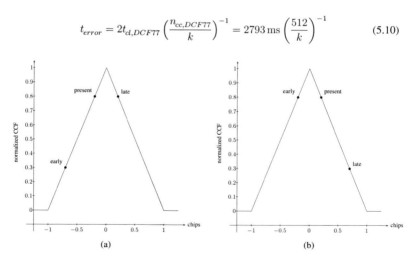

Figure 5.11: Different possibilities of false maximum definition of DCF77 CCF

In case no correction algorithm is applied to the calculated EPL data structure, an additional error to the time offset measurement has to be added. In the worst case, this error can have a size slightly less than the distance between channels x and $x + 2$ in the correlation bank. Uncertainty in time offset measurement in that magnitude would not allow to precalculate the code phases for the GNSS signals with sufficient precision such that a limitation od the search space is possible. Fortunately, as the shape of the CCF result is theoretically known, the results from the baseband processing can be fed into an interpolation algorithm calculating the global maximum. The present architecture corrects the calculated code phase of the DCF77 signal by applying a linear interpolation based on the EPL values gathered from the baseband signal processing.

Figure 5.12 shows two different cases for calculated EPL values where the present path does not match the global maximum of the CCF. This can easily be seen from the fact, that early and late path results are not equal. The algorithm computes the correct position of

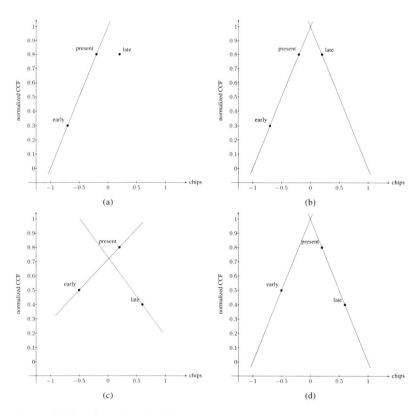

Figure 5.12: Linear interpolator for CCF maximum: (a) Straight line through early and present points, (b) Intersection calculation, (c) Failure of exact code phase calculation, (d) Straight line through present and late points first

the CCF maximum by applying the EPL values to two linear equations. The intersection of both straight lines delivers the corrected maximum $nCCF_{cor}$ of the CCF as given in figures 5.12(a) and 5.12(b). First, the early and present values are used to define and calibrate the straight line equation to run through these points. Second, another straight line with inverted slope is calculated running through the point determined by the late correlator channel.

In case the calculated intersection point does not lie between the early and late points on the x-axis and is not greater or equal to the CCF value of the present path, the computation of the first linear equation is restarted with the present and late points as demonstrated in figures 5.12(c) and 5.12(b).

A second possibility to interpolate the real maximum value of the cross correlation functions is to apply a sinc-interpolation according to the Whittaker-Shannon interpolation formula [63]. Due to the nature of the algorithm, working with a higher sampling rate, it is possible to achieve results with much higher accuracy, compared to a linear approach. Similar to the calculation of the cross correlation function again there is the possibility to either perform the sinc-interpolation in time or in frequency domain. This offers the flexibility to apply the same interpolation scheme to frequency domain based calculations of the CCF as well as to the time domain based approaches. To save implementation resources it is preferable to stay in the domain chosen for the calculation of the CCF, here in the time-domain. The algorithm to be applied is given in equation (5.11):

$$
\begin{aligned}
nCCF_{cor} &= max\{\mathbf{nCCF_{uip}}\} \\
\mathbf{nCCF_{uip}} &= \sum_{p=1}^{m} \mathbf{nCCF_{uncor}}\, sinc(f_{uncor}\,(k - \mathbf{t_{uncor}})))
\end{aligned} \tag{5.11}
$$

The corrected maximum $nCCF_{cor}$ of the results of the non-coherent CCF is represented by the maximum of the upsampled and interpolated series $nCCF_{uip}$. Each element of this series is calculated by summing the results of the multiplication between the uncorrected results from the non-coherent CCF $\mathbf{nCCF_{uncor}}$ with a sinc function. The argument of the sinc function is derived from the multiplication of the sampling frequency for the correlation channels output vector f_{uncor} with the difference between the sample value corresponding to the current upsampled and interpolated results k and the sampling times of the CCF channels output $\mathbf{t_{uncor}}$.

In any case the corrected DCF77 code phase can be applied to equation (5.12) resulting in the correct offset between local time and UTC.

$$
t_{offset} = t_{chip,DCF77}\, nCCF_{cor,DCF77} = 793\,\mathrm{ms}\,\left(\frac{512}{k}\right)^{-1} nCCF_{cor,DCF77} \tag{5.12}
$$

Finally the corrected code phase is made available to the host platform.

Once precise information about reception time has been gathered, the algorithm starts the positioning operation. The basis for the positioning operation is the TOF measurement to different SVs. This approach is comparable to common architectures since the enhancements introduced by the present architecture only concern resource usage but does not incorporate a different principle of operation. Therefore, it is still necessary to have access to the actual navigation message although it is not decoded from the satellite signal itself. The ephemeris data is made available by download from a reference station via the cellular network connection.

Common AGNSS architectures gather their assistance data via standardized assistance protocols over cellular networks, as defined for example by the 3rd Generation Partnership Project (3GPP) [34]. To provide GNSS assistance data the base stations have to be equipped with a Location Measurement Unit (LMU). Although the protocols are standardized, unfortunately still not all base stations are or will be equipped with a LMU. Regarding networks like Wireless Local Area Networks (WLANs) assistance protocols are not specified anyway.

Therefore, the present architecture follows a different approach. It makes use of Navigation Message Files in the Receiver Independent Exchange Format (RINEX) [64] which are available for many reference stations around the world. These files are constantly provided via File Transfer Protocol (FTP) servers through the internet. Data is normally updated every hour and available for download free of royalty. The RINEX Navigation Message File contains the navigation message data for the satellites in view at the appropriate reference station as described in section 2.2.2. Together with the corrected system time the receiver software is now capable of determining the satellites in view and their position taking receiver's approximate position as a basis together with the corrected system time. Although most of the time a higher number of satellites are visible to the device, only four are needed to solve the system of equations (2.4) introduced in section 2.1.1.

After choosing the appropriate SVs for the positioning operation the receiver has to determine the approximate code phase for each satellite to be used for the positioning operation with respect to approximate position and time. The distances to the SVs are calculated and converted to the number of chips by equation (5.13) using the approximated user position $\{x_{user,approx}, y_{user,approx}, z_{user,approx}\}$ the calculated satellite position position $\{x_{sat,j}, y_{sat,j}, z_{sat,j}\}$, the number of chips per code cycle n_{cc} and the channel spacing k.

$$
n_{approx,j} = \left(\left[\sqrt{(x_{user,approx} - x_{sat,j})^2 + (y_{user,approx} - y_{sat,j})^2 + (z_{user,approx} - z_{sat,j})^2} \right] \right.
$$
$$
\left. \mathrm{mod} \left[299792458 \, \frac{\mathrm{m}}{\mathrm{s}} 10^{-3} \, \mathrm{s} \right] \right) \left(299792458 \, \frac{\mathrm{m}}{\mathrm{s}} 10^{-3} \, \mathrm{s} \frac{k}{n_{cc,GNSS}} \right)^{-1}
$$

$$(5.13)$$

Additionally the number of complete code cycles $n_{epochs,j}$ according to equation (5.14) has to be computed and stored for later processing.

$$n_{epochs,j} = \left\lfloor \left(\left(\left[\sqrt{(x_{user,approx} - x_{sat,j})^2 + (y_{user,approx} - y_{sat,j})^2 + (z_{user,approx} - z_{sat,j})^2} \right] \right. \right. \right.$$
$$\left. \left. \left. \div \left[299792458 \, \frac{\text{m}}{\text{s}} 10^{-3} \, \text{s} \right] \right) \left(299792458 \, \frac{\text{m}}{\text{s}} 10^{-3} \, \text{s} \frac{k}{n_{cc,GNSS}} \right)^{-1} \right) \right\rfloor$$

$$(5.14)$$

After the precalculation of the expectable code phases, the record of data controlling the digital baseband processing can be assembled and transferred. The dataset for each SV comprises the approximated initial code phase for satellite j, $n_{approx,j}$, the associated SV number as well as the appropriate Doppler shift frequency. The latter two can directly be derived from the RINEX navigation file. The information extracted from the data file, especially regarding Doppler frequency, is only valid if the user position and the position of the reference station do not differ extensively. Together with increasing distance of user position and reference station, the estimation of SVs in view as well as their specific Doppler frequency deteriorate.

The baseband signal processing unit, now reused to calculate the GNSS signal code phases, returns the uncorrected outcome of the non-coherent cross correlation function between the satellite signal and the local generated replica. Comparable to the situation when generating precise time information from the DCF77 signal, the calculated maxima at the output of the non-coherent CCF do not necessarily reflect the real maxima, due to the finite distance between adjacent CCF channels. The reuse of either the linear interpolation or sinc interpolation architecture is possible. The appropriate architecture is applied to the results of the cross correlation functions between the GNSS signals and their local generated replicas.

Once corrected code phases nCCF$_{cor}$ are known, the conversion to distances d$_{u-sv,j}$ between user and satellites can be performed applying equation (5.15).

$$d_{u-sv,j} = c \left(n_{epochs,j} \, t_{epoch} + nCCF_{cor} \, t_{chip} \right) \tag{5.15}$$

The results from this function are deteriorated due to the influence of delays introduced by several atmospheric layers. These effects were not taken into consideration to up to now. There are two layers which account for a significant modification of the satellite signal, the ionosphere and the troposphere.

Although there are several models calculating the ionospheric influence the presented architecture uses the Klobuchar model as described in section 2.1.3.

The WAAS model, described in section 2.1.3, is used to determine the delay introduced by the troposphere. In contrast to the ionospheric delay, which can be estimated by equation (2.7), the tropospheric delay has to be considered in the iterative positioning solution. Therefore an initial solution based on equations (5.16) is calculated.

$$\sqrt{(x_{sat1} - x_{user})^2 + (y_{sat1} - y_{user})^2 + (z_{sat1} - z_{user})^2} = d_{u-sv,j} + c\, t_{error}$$
$$\sqrt{(x_{sat2} - x_{user})^2 + (y_{sat2} - y_{user})^2 + (z_{sat2} - z_{user})^2} = d_{u-sv,j} + c\, t_{error}$$
$$\sqrt{(x_{sat3} - x_{user})^2 + (y_{sat3} - y_{user})^2 + (z_{sat3} - z_{user})^2} = d_{u-sv,j} + c\, t_{error}$$
$$\sqrt{(x_{sat4} - x_{user})^2 + (y_{sat4} - y_{user})^2 + (z_{sat4} - z_{user})^2} = d_{u-sv,j} + c\, t_{error}$$

$$(5.16)$$

In a second step the delays introduced to the SV signals by the troposphere are calculated and the pseudorange for the j-th SV determined by the initial iteration of the closed form navigation solution is corrected according to equation (5.17)

$$d_{u-sv,j} = d_{u-sv,j} - \left((t_{zenithdrydelay} + t_{zenithwetdelay})\, c \right)$$

$$(5.17)$$

The parameters $t_{zenithdrydelay}$ and $t_{zenithwetdelay}$ are calculated depending on the model defined in [65]. The model mainly depends on geodetic latitude, height and the day of year, which makes clear that an initial positioning solution must be available to the troposphere correction model. It is necessary to use the results of the first iteration of the closed form navigation solution instead of the initial position estimation based on the cellular network connection. False correction parameters result from significantly different ranges to the SVs calculated for both solutions.

The iterative closed form solution can be based on any solution proposed in literature such as [66] [67].

CHAPTER 6

Systemsimulation

In order to prove operational capability and determine key parameters of the architecture introduced in chapter 5 a system simulation has been set up. Since even high performance computing is not able to perform a full system simulation especially including RF and baseband signal processing in full bandwidth all simulations are performed in the Equivalent Complex Baseband (ECB) without loss of generality. Effects degrading the GNSS signal quality have to be transferred into an equivalent ECB model for proper simulation .

6.1 Simulation Setup

The simulation model used in the present work simulates the DCF77 as well as GPS satellite signal at a virtual mobile device. For proper simulation several key parameters must be present to allow a realistic simulation of the incoming signals:

- Current time

- GPS ephemeris data

- Current user position

- Estimated current user position

- DCF77 transmitter position.

Using current user position, current time and the DCF77 transmitter position, the DCF77 signal at the receivers antenna can be simulated. The dependencies and characteristics of the DCF77 signal model are investigated in section 6.1.1.

Again using current time and user position together with the GPS ephemeris data, the satellite constellation can be predicted and satellite signals arriving at the user position can be simulated. The model implemented to generate the satellite information is presented in 6.1.2.

Since both signal models reflect their output in the ECB, limitations and distortions introduced by RF Frontends are included or added to the channel model. The signal processing in the receiver is modeled in 6.1.5, implementing the ideas and the apporach of the architecture given in 5.

6.1.1   DCF77 SIS Model

The DCF77 SIS model should reflect the generation of the DCF77 signal at the transmitter site as well as its emission by the antenna and the propagation to the receiving antenna. While generation of the information to be transmitted is discussed in depth in literature [5] [8], parameters regarding the emission of the signal cannot that easily be derived. Although the impact of the transmitting antenna on the SIS is discussed by Hetzel [5] only qualitative statements are made, especially regarding the impact on the phase of the transmitted signal. However, remembering the results presented in [5] the influence of the transmission antenna must not serve as an important error source. Nevertheless, keeping the in-depth discussion on DCF77 SIS in chapter 4 in mind, three dominant sources influencing signal quality and information content of the DCF77 SIS at the receiver's antenna can be identified, namely:

• Step response of the transmitting antenna

• Radio channel, especially in-band interference transmitters near the receiver

• Distance transmitter-receiver.

Not only the influence of these factors has to be covered by an appropriate channel model, but it must transfer the distortions introduced to the ECB for proper modelling.

The effects on the transmitted signal based on antenna imperfections have been investigated by Hetzel [68]. The influence of the transmitting antenna on the SIS is mainly determined by its limited ability to follow the steering signal, hence the modulation of the RF-signal. This is due to the massive capacitive top loading of the antenna, representing together with the antenna itself a resonant circuit which still suffers hard to follow the implemented modulation schemes. The limited bandwidth of the antenna results in a scattered step response in phase and amplitude, independent of which modulation scheme changes its value (AM or PSK). Of course it is impossible to characterize the antenna in terms of measuring S-parameters and build an antenna model due to the large proportions. Instead the influence of the transmitting antenna must be indirectly modeled by applying a filter characteristic to the signal model, generating an output spectrum similar to the one described in literature. Here again the work of Hetzel must be cited [5] and the resulting spectrum should be comparable to the those given in Figure 4.2. Additionally the model must shift the spectra of the SIS to the ECB.

Ignoring imperfections introduced by the surrounding (e.g. soil moisture) or weather conditions (humidity, height of ionospheric D-layer), the antenna represents a small-bandwidth bandpass filter formed by a LC-resonant circuit. The conversion of the RF-bandpass characteristic to the ECB results in a shift of the output spectrum to be centered at 0 Hz and the appliance of an adjacent low-pass filter. The bandpass filter characteristic used in this work, trying to fit the spectrum of the transmitted signal of at the virtual DCF77 site according to [8] is given in figure 6.1.

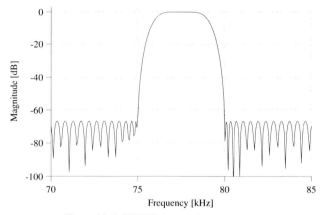

Figure 6.1: DCF77 TX antenna frequency response

The channel model used to describe the influences disturbing the traveling RF signal is based on an Additive White Gaussian Noise (AWGN) channel. This channel model can easily be expanded to the second error source, in-band interference. Two main scenarios are imaginable, degrading sensitivity to an extent such that restoration of the BPSK information is not possible anymore. The usage of a PRN as modulation sequence makes the transmission insensitive to in-band interference as long as the interferer makes use of orthogonal modulation sequences. But of course this is only true as long as sufficient $\frac{C}{N_0}$ can be achieved, too.

The first scenario, a PRN modulated interferer in the frequency band of the DCF77 service, is unlikely to occur since the frequency allocation by the Bundesnetzagentur [69], does not allow the usage of parallel services in the same band.

The second scenario instead is realistic. Although there is no service using the same band, parasitic harmonics of other electro-magnetic fields can overdrive the DCF77 receiver's RF frontend (for example the 5th overtone of the line repeating frequency of Cathode Ray Tube (CRT) monitors at 78.125 kHz). To model this effect a Continuous Wave (CW) signal at the interfering frequency can be added to the simulated incoming DCF77 signal, reducing the $\frac{C}{N_0}$ appropriately.

The last error source to be modeled is the distance between transmitter and receiver with feedback on the radio channel model. It has two main effects on the signal that has to be taken care of:

- the power level of the incoming signal, mainly determined by the channel model,

- the phase shift of the information due to the travel time of the signal.

As stated earlier an AWGN channel model is used for the simulation of the DCF77 signal in this work. Although an AWGN channel model is not ideal suitable for terrestrial radio communication, it is sufficient for modeling the reception of the DCF77 signal at exposed, hence nearly LOS positions, where effects like multipath propagation, dispersion or shadowing do not play an important role. Under the idealized assumption of an AWGN channel the attenuation K_{DCF77} of the DCF77 signal can then be estimated according to Friis transmission equation (6.1) [70].

$$K_{DCF77}[dB] = G_t G_r \left(\frac{\lambda}{4\pi R} \right)^2 \tag{6.1}$$

The resulting attenuation K_{DCF77} measured in dB is dependent on both transmission and receiving antenna gains, G_t and G_r respectively, as well as the wavelength λ and the distance between transmitter and receiver R. According to [8] the DCF77 antenna is fed with 50 kW power, of which approximately 30 kW to 35 kW [8] are radiated effectively. To guarantee effectiveness even in bad signal reception conditions the lower boundary of radiation power is assumed for all simulations below. Together with a typically antenna gain of the receiving antenna of 0-6 dB [71] the resulting received signal strength can be determined by equation 6.2. Not only the geometric dimensions (diameter and length) and component values of the ferrite rod antenna are important for the effective antenna gain but orientation with respect to the transmitter, too. Here the same arguments take effect and it is legal to assume 0 dB of antenna gain for worst-case assumption, leading to:

$$S_{RX\_DCF77}(t) = \sqrt{P_{TX\_DCF77} 50\,\Omega} S_{TX\_DCF77}(t) G_r \left(\frac{\lambda}{4\pi R} \right)^2$$

$$= \sqrt{P_{TX\_DCF77} 50\,\Omega} S_{TX\_DCF77}(t) \left(\frac{\lambda}{4\pi R} \right)^2 \tag{6.2}$$

Equation 6.2, a modified Friis transmission equation [70], only delivers valid reuslts for the calculation the signal strength of the ground wave of the DCF77 signal. For a full coverage a second signal source covering the backscattered signal from the atmosphere must be modeled and added to S_{RX_DCF77}, too. Since at distances below 1000 km from the transmitter, signal strength of the ionospheric wave is orders of magnitude below the power of the ground wave, modeling the scattered wave is omitted here. The channel model used herein is expanded to cover extra attenuation due to bad reception circumstances, e.g. due to shadowing. The additional deterioration model of the DCF77 signal is here limited to signal strength. Other effects, e.g. multipath propagation are omitted due to insignificance. This constraint is acceptable since shadowing is the main influence on the DCF77 signal, besides cancellation due to overlap of ground- and atmospheric wave.

Besides applying equation (6.2) to the transmitted signal S_{TX_DCF77} for modeling signal attenuation, modification effects regarding the encoded information in the transmitted signal have to be considered. Main influence on the DCF77 information is the phase shift caused by the travel time of the RF wave, which has to be taken into account during simulation. Converted to the ECB the distance between transmitter and receiver and consequently the Time-of-Flight, corresponds to an appropriate phase shift of the DCF77 PRN sequence. The resulting phase shift n_{tof} is derived from equation (6.3) using the time of flight t_{tof} and the length of one symbol. This again can be derived from the number of symbols per epoch N_{PRN} divided by the length of one epoch t_{PRN}. The TOF is calculated by dividing the distance $d_{tx\text{-}u}$ transmitter to user by the speed of light c.

$$
\begin{aligned}
n_{tof} &= t_{tof} \frac{N_{PRN}}{t_{PRN}} = \frac{d_{tx-u}}{c} \frac{N_{PRN}}{t_{PRN}} \\
&= \frac{d_{tx-u}}{c} \frac{645.83 \, \text{chips}}{1 \, \text{s}}
\end{aligned}
\tag{6.3}
$$

Additionally the phase shift of the DCF77 PRN sequence has to be modified to represent current time, too. Therefore equation 6.3 has to be extended by the time interval between current time t_{abs} and current epoch t_{epoch} as given in equation 6.4.

$$
n = (t_{tof} + t_{abs} + t_{epoch}) \frac{N_{PRN}}{t_{PRN}}
\tag{6.4}
$$

By applying an upsampling factor of 10000 to the simulated DCF77 SIS prior to the adaption of the PRN phase shift by n allows an sufficient accurate mapping of the phase shift to the PRN sequence to fullfill the Nyquist-Shannon theorem. The result from equation (6.4) real number n needs to be modified and truncated adequate.

Again it has to be emphasized that the present model for the DCF77 SIS is only applicable to situations when the ground wave delivers the dominant portion to signal strength received by the antenna. Hence trying to simulate positions at greater distances than approximately 1000 km from the transmitter of the DCF77 signal at Mainflingen, Germany, where signal strength of the ionospheric wave dominates, will not deliver significant results.

6.1.2   GNSS SIS Model

The GNSS SIS arriving at the receiver antenna is the superposition of the individual signals of the SVs in view and is certainly modeled accordingly. Each individual satellite signal again is influenced by five factors:

- Noise and attenuation

- PRN phase-shift

- Doppler frequency

- Movement of satellite during reception interval

- Absolute time

Similar to the simulation of the DCF77 SIS under the assumption of good reception conditions at the user position, hence a clear view to the sky, effects like shadowing or multipath propagation of the GNSS SIS can be omitted and an AWGN channel model is suitable. Consequently this approach determines noise and attenuation modelling. The user-satellite distance additionally is responsible for the phase shift of the information encoded in the GNSS SIS. The resulting incoming signal to the receiver is modeled as a superposition of the individual signals of the SVs in view superimposed by the thermal noise level. Compared to DCF77 or other transmission beacons, the main difference when simulating GNSS SISs is the motion of the SV during signal transmission. The movement of the satellite with respect to the user position results in an individual Doppler frequency shift of the carrier frequency for each SV. Consequently the position of the SV is changing during reception time, which has to be covered by the channel model, too. Similar to the DCF77 SIS simulation, information encoded in the SVs signal is influenced by the absolute time resulting in an appropriate phase shift of the PRN.

To simulative cover the mentioned effects, calculation of GNSS satellite positions at transmission time is necessary at first, subsequently allowing the TOF determination of the GNSS signal between user and SV. Second, due to the relative speed between user and SV, of slightly below $14000 \frac{km}{h}$ in average [60], the residual Doppler frequency offset can be identified. Realistic target operation scenarios for mobile devices, where velocity of the user is negligible, allow the assumption of SV velocity as relative speed between user and SV. Depending on the elevation angle of the satellite above horizon at user's position and the direction of the satellite's movement vector, a frequency offset of ±5 kHz from the nominal carrier frequency 1.57542 GHz has to be considered. Fortunately satellite positions and consequently their movement can be derived using the appropriate ephemeris data and current time. Several internet services make GNSS ephemeris data available free of charge, closely coupled to the information actually transferred by the real satellite signals. In this work the model relies on data provided by The Crustal Dynamics Data Information System (CDDIS) [72] from the Goddard Space Flight Center of the NASA. Solving equations 2.8 introduced in chapter 2 using the ephemeris data together with absolute current time leads to SV positions, already compensated by ionospheric and relativistic effects. Consequently the two main error sources disturbing the GNSS satellite signal do not need further attention during modeling. Less severe error sources such as tropospheric effects which do directly concern the GNSS SIS and can be modeled by an adapted satellite position are not included in this work. The legality af this approach can easily be approved, considering the low impact on the position solution of only ±0.5 m [13].

The effective phase shift to be applied to the SV's PRN sequence is composed of two fractions, first the phase shift arising from signal's TOF and a second the part due to current time. Assuming precise satellite position solutions are available, calculation of the appropriate phase shift for each satellite's TOF respectively, can be achieved by the algorithm in equation 6.5, using Gauß bracket notation [73].

$$
\begin{aligned}
d_{chip} &= \frac{d_{code\_epoch}}{n_{PRN}} = \frac{c\, t_{code\_epoch}}{n_{PRN}} \approx \frac{299792458\,\frac{m}{s}\; 1\,ms}{1023} \approx 0.293\,m \\
n_{full\_epochs} &= \left\lfloor \frac{d_{sat\_user}}{d_{code\_epoch}} \right\rfloor \\
d_{residual} &= d_{sat-user} - \left(n_{full\_epochs}\, d_{code\_epoch} \right) \\
m_{tof} &= \frac{d_{residual}}{d_{chip}}
\end{aligned}
\tag{6.5}
$$

The length of a NAVSTAR GPS epoch of 1 ms, is converted to its distance equivalent d_{code_epoch} by multiplication with the speed of light c. Correspondingly, the distance per PRN chip d_{chip} results from the division of d_{code_epoch} by the number of chips per epoch n_{PRN}. Since PRN shifts due to full code epochs n_{full_epochs} along the distance between user and satellite $d_{sat-user}$ are redundant during generation of the simulated signal, only the residual distance $d_{residual}$ after subtracting the full code epochs are of interest. The final number of chips m_{tof} the PRN sequence must be shifted due to the TOF can then be calculated.

When adapting the SIS model to current time, a prior look at the SIS structure, repeating the appropriate PRN sequence every millisecond, immediately shows that only the residual non-full code epoch shift is significant for modelling and redundant full code epoch shifts can be dropped. Hence the length of full code epochs is subtracted from the current time, leading to the period since the last millisecond mark. The resulting number of chips m_{time} the GNSS PRN sequence must be shifted due to current time t_{curr} can therefore be calculated from equation 6.6.

$$
\begin{aligned}
t_{res} &= t_{curr} - \lfloor t_{curr} \rfloor \\
m_{time} &= \frac{n_{PRN}}{t_{code\_epoch}}\, t_{res}
\end{aligned}
\tag{6.6}
$$

The overall resulting number of samples m_{res} the PRN sequence must be shifted for each satellite in view individually results from the superposition of m_{time} and m_{tof}.

While absolute time as well as TOF affect the information encoded in the satellite signal, the distance between SV and user has two additional impacts on the GNSS SIS, too. Again Friis formula [70] comes into play, modeling signal attenuation in the RF-channel. Together with the transmit power of 21.9 W on the L1-signal and an transmit antenna gain of 13 dBi [74] the attenuation of the NAVSTAR GPS signal yields to equation (6.7).

$$K_{GPS} = \sqrt{P_{TX\_GPS}50\,\Omega}\,G_t\,G_r\left(\frac{\lambda}{4\pi R}\right)^2$$

$$= \sqrt{21.9\,\text{W}\,50\,\Omega}\,13\,\text{dBi}\,3\,\text{dBi}\left(\frac{\lambda}{4\pi R}\right)^2 \quad (6.7)$$

Second the movement of the satellites introduces a Doppler shift of the carrier frequency of ±5 kHz. Although the effect of the Doppler shift on the incoming signal is generated during TOF, its influence is investigated in 6.1.3 due to its appearance based on the RF-to baseband down conversion process in the RF-Frontend of the receiver. Combining the NAVSTAR GPS signal generation model with the GNSS channel model for all individual SVs, yields to $S_{SIS,GPS}(t)$ according to equation (6.8).

$$S_{SIS,GPS}(t) = \sum_{k=1}^{n} K_{GPS,k}\,S_{TX\_GPS,k}$$

$$= \sum_{k=1}^{n} \sqrt{P_{TX\_GPS}50\,\Omega}\,S_{TX\_GPS}(t)\left(\frac{\lambda}{4\pi R}\right)^2 \quad (6.8)$$

$$\left(\sin(2\pi f_{Doppler})\,PRN_{sat,k}(t - m_{res,k})\right)$$

The effective GNSS SIS as it appears at the antenna of the receiver is composed from the superposition of the individual SV signals. These again are depending on the attenuation factor and the information shift as well as the Doppler frequency as introduced earlier.

6.1.3   DCF77 RF Frontend Model

The imprecisions inherent to the architecture of the DCF77 analog-frontend as illustrated in Fig. 5.2 in section 5.2.2 also have significant influence on result of the baseband signal processing engine. To model the DCF77 frontend a superposition of individual transfer functions for the components of the RF-frontend, as given in equation 6.9, is applied to the DCF77 SIS. Where mandatory the real transfer function has to be modified to describe the influence of the specific component in the ECB.

$$S_{DCC77} = \left(\left(S_{DCF\_SIS}H_{ant}H_{amp}H_{ADC}\right)H_{mixer}\right)H_{LP} \quad (6.9)$$

The response function H_{ant} models the limited bandwidth of the antenna in the current system. The antenna, built up from a LC-resonant circuit whose q-factor is improved by a ferrite rod, is modeled by a band-pass filter, as illustrated by the red graph in Fig. 6.2. The antenna characteristic must be designed to provide sufficient bandwidth for the reception of the PSK modulation, according to the spectrum of the measured transmitted signal in Figure 4.2. As described earlier in chapter 5.2.2 the information generated after demodulation of the PSK is the exact determination of the SIS phase, hence signal's zero crossing. Therefore it is sufficient to have an antenna passband covering the fundamental and the first overtone of the modulation spectrum. Converting the antenna model to the

ECB results in a shift of the transfer function to baseband, resulting in the blue graph in Fig. 6.2. Noise introduced by the components is modeled by a cumulative noise figure for the receiver which is handled later. The cumulative noise figure can be derived by the applying Friis formula [75].

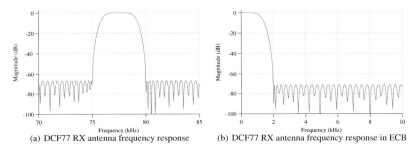

(a) DCF77 RX antenna frequency response   (b) DCF77 RX antenna frequency response in ECB

Figure 6.2: Original and converted DCF77 RX antenna model

Since signal strength at the antenna maybe very weak depending on the distance between receiver and transmitter amplification of the incoming signal is necessary in the receiver. The voltage drop over the antenna in rare cases exceeds several μV. Therefore the amplification of the incoming signal is necessary to feed the demodulating mixer correctly. The applied amplification stage is modeled by the transfer function H_{amp}. Common architectures for DCF77 receivers, who only evaluate the information encoded in the AM, must assure both, not to overdrive the amplifier to regain the amplitude information and guarantee linearity of the amplification process. The herein proposed architecture for a DCF77 receiver, as mentioned in chapter 5.2.2 instead intentionally overdrives the amplification stage, generating a square wave signal at its output. As long as the input signal to the amplifier has no DC-component, the information fed to the following mixer stage is only dependent on the phase of the DCF77 SIS. Accordingly the transformation of the amplifier characteristic to the ECB is realized by shifting the frequency response of the amplifier to baseband. Besides noise introduced by the amplifier, which is as stated earlier, already covered by the cumulative noise figure of the whole receiver a simple multiplication of the signal amplitude can be assumed as the amplifier model. Of course this assumption is only legal if the frequency response of the amplifier is linear in the DCF77 RF frequency bandwidth and compensated for other effects like e.g. temperature. Also neither center frequency of 77.5 kHz nor bandwidth comprising 3 kHz deliver special demands to the amplifier circuit. Since there are circuit designs known insensitive to environmental parameters as well as showing sufficient linearity, ideal conditions can be supposed fot the applied model here. Consequently the assumption of modeling amplification with a simple amplitude multiplication and limiting the output value to the maximum of the input of the ADC model is legal.

The amplified and amplitude limited signal delivered by the amplifier model is fed into the ADC model H_{ADC}. Since only the phase of the incoming signal is of interest here, the zero crossing of the incoming signal is the must be detected very precise. Therefore it is more desirable to achieve a high sampling rate instead of higher resolution particularly because the amplitude information is erased by the overdrive operation of the preceding amplifier. However when comparing both graphs in Fig. 6.3 it is obvious that the implementation of a 1 bit ADC is sufficient to determine the correct code phase, without loss of accuracy in the results of the CCF.

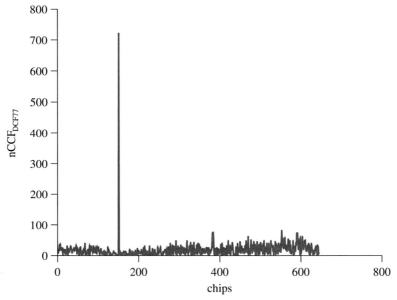

Figure 6.3: Results of the CCF for DCF77 signal simulation implementing either 1 bit(blue) and 2 bit(red) ADCs

Main error source to be modeled at this point in the receiver chain is the jitter of the ADC sampling clock. Assuming a sampling frequency, way above the frequency defined by the Shannon-Nyquist criterion (here a sampling frequency of at least 158 kHz assuming a band filter width of 3 kHz), still sampling rates at several MHz are not exceeded. Provided sufficient stability of the clock generating circuit, which is not a challenge nowadays in the desired band of several MHz, possible errors introduced by the transition to the digital domain can be omited. Consequently the sampling process can be assumed to be implemented in an 1-bit ADC or a simple flip-flop circuit, while residual additive jitter as far as existent is modeled at the LO signals of the following mixer stage.

The demodulator, responsible for downconversion to baseband and demodulation of the DCF77 signal whose influence is modeled by H_{mixer} in equation 6.9 can be thought of a simple multiplication together with an adjacent filter. However, the simulation of the mixer is obsolete since data already is available in the ECB such that only parasitics have to be overlooked. According to Fig. 5.2 this operation is already performed in the digital domain. Consequently effects like truncation or similar are not necessary to be modeled for proper operation.

Although functionality of the the mixer is omitted while simulating in the ECB, the low pass filter removing unwanted mixing products is available in reality and has consequences on the resulting data. Its influence is described by H_{LP} in equation 6.9. The filter model does not have to be adapted to the ECB since all operations following the mixer stage are already performed in baseband. The lowpass filter resided at the output of the mixing stage is implemented as symmetric Finite-Impulse-Response (FIR) filter structure realizing the filter characteristic given in Fig. 6.4. Additionally a separate RF signal simulation shows significant attenuation of the mixing products in the spectrum. They form a peak at 155 kHz in the red graph of Figure 6.4, which is significantly reduced after applying the filter characteristic illustrated by the blue graph, while not affecting the output represented by the green graph.

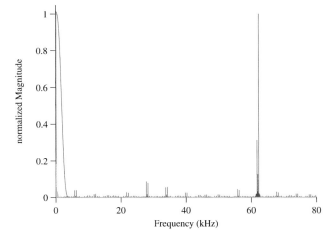

Figure 6.4: Mixer output (red) compared to Filter output (green) (overlapping red graph) filtered with by lowpass characteristic (blue)

The baseband spectrum of the input (red) as well as output (green) signals of the filter removing the unwanted frequency parts of the demodulation process response have been zoomed to prove undisturbed transfer of the Phase Modulation (PM) encoded information as illustrated in Figure 6.5.

Main error source immanent to the information output of the DCF77 RF of the mixer and filter stage in the DCF77 RF Frontend is a phase delay, dependent on the filter charateristic. Regarding the filter it can analytically derived from the filter generation algorithm (231.5 ° in the available architecture). Regarding the mixer stage it is dependent on the clock frequency of the baseband processing engine. Instead the phase shift due to the ana-

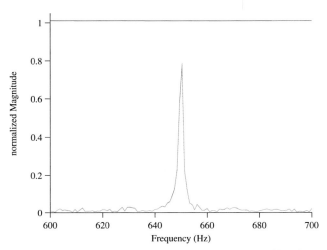

Figure 6.5: Mixer output (red) compared to Filter output (green) filtered with by lowpass characteristic (blue), zoomed to highlight baseband output

log RF Frontend can be empirically determined by measurement. Compensation of the overall phase shift is handled during determination of time information in the algorithms later.

6.1.4   GNSS RF Frontend Model

For sake of simplicity the GNSS RF Frontend model does not cover all effects introduced to the GNSS signal. Besides noise and phase shift, second order effects like IQ-impairments or clock stability of the ADC sampling clock are not modeled, assuming sufficient $\frac{C}{N_0}$ at the digital output of the GNSS RF Frontend. According to Ehm [32] a 2 bit quantization is optimal for the digitization of the GNSS signal. Therefore the ADC is modeled this way.

Based on equation (6.8) and ignoring additional deteriorating effects, the GNSS signal at the downconversion mixer input can be described by equation (6.10). It is virtually filtered by the antenna's frequency response as well as an adjacent band filter. Both characteristics have to be selective such that the frequency band of the GNSS is not affected. Under the assumption of good reception conditions still two parasitic effects remain. The first is attenuation due to the insertion loss of the band-limiting filter as well as an imperfect bandpass characteristic of antenna and filter. Both are ignored, since they develop severe impact to the positioning solution only under bad reception conditions and commercial solutions are available delivering the desired performance for the reception of GNSS signals. The absolute value for $\frac{C}{N_0}$ when the navigation solution collapses is of minor interest in the prove of concept for the architecture at hand.

Second a phase shift of the received and demodulated signal due to group delay in the RF Frontend components is introduced and has to be modeled. Comparable to the processing of the DCF77 signal in its analog Frontend the remaining phase shift can and must be compensated in the digital baseband processing and is therefore handled and discussed there.

$$
\begin{aligned}
t_{res} &= A_{att}\, sin(\omega t + \phi(t)) sin(\omega t + f_{doppler} t) \\
&= A_{att}\, sin(\omega t - \omega t + \phi(t) + f_{doppler} t) + A_{att}\, sin(\omega t + \omega t + \phi(t) + f_{doppler} t) \\
&= A_{att}\, sin(\phi(t) + f_{doppler} t)
\end{aligned}
$$
$$(6.10)$$

Applying a low pass filter adjacent to the micer process, eliminating the high frequency parts of the down-converted signals delivers the ECB baseband signal according to equation 6.11.

$$
d_{gnss} = A_{att}\, sin(\phi(t) + f_{doppler} t) \tag{6.11}
$$

6.1.5 Digital Baseband Signal Processing Model

Inputs to the baseband signal processing are the received, filtered and demodulated DCF77 or GNSS signals as well as approximate time and position information. The model for the baseband signal processing is congruent to its realization, ignoring the hardware/software boundary and its implication on the determination of the positioning solution (either serial or parallel realization depending on available hardware resources). This approach is suitable since besides time delay in the transfer of data between hard- and software, no deterioration of the data occurs neither in the model nor in reality.

The main algorithms performed in the digital baseband processing are already introduced in chapter 5.3.3. Prior to the cross correlation operation according to approximate local time and position the local PRN replicas are generated. Therefore the appropriate PRN sequence together with Doppler data in case of GNSS signals, are loaded from a Look-up-Table (LUT) modeling the ROM block of the architecture. Especially when performing GNSS data it is essential to calculate the satellites in view and their positions at current time. The calculated positions are subsequent corrected due to their movement during the time length of reception. Afterwards the correlation is calculated according to the equations in chapter 5.3. Finally the results generated from the cross correlation function have to be interpolated due to the limited sampling frequency of the reciever.

6.1.6 Positioning Solution

The positioning solution either modeled or realized reflects the equations introduced in section 2.1.1 for calculating the current position from the SV observations, namely the pseudoranges to the SVs. Both model and realization do not differ, hence the algorithm described in equations 2.8 can be adopted directly and be realized in software.

Two possible issues may be overlooked when implementing equations 2.8. First the calculation of the positioning solution must be realized in a way such that the algorithm has been finished until the next positioning request is made to the receiver. This demand can mainly be answered by processing power, which is sufficiently available even in embedded systems. Otherwise dedicated hardware can be spent to overcome possible issues. Second the definition of the internal variables of the processing unit must deliver sufficient resolution for solving all equations to an adequte precision, better than the resolution delivered by the information content of the receiver. Possible issues are imposed by the limitations introduced by the word width of the processor architecture. However, modern (embedded) processor architectures realize 64-bit word widths, delivering precision way beyond the possibilities of the received signals.

Therefore no special model for the realization of the positioning algorithm has to be developed instead a straight forward implementation of equations 2.8 can be used.

6.2   Simulation Results

This section illustrates outputs of the system simulation environment discussed in the preceding sections. Focus is to demonstrate the behavior of the DCF77 and GNSS SISs under different reception conditions and identify potential sources for errors in the algorithm determining the current position and their impact on the positioning solution.

First a closer look on the results of a stand-alone DCF77 signal simulation is thrown, trying to carve out the influence of timing as well as position approximation errors on the timing information generated. Second the results of the GNSS signal simulation and positioning operation are highlighted incorporating the given results of the timing simulation.

6.2.1   DCF77 Simulation Results

As described in-depth during the preceding chapters superiority of the proposed architecture compared to common architectures is gained by the knowledge about reception time. Simulation results in the following section will highlight the potential limitations of time synchronization due to reception conditions as well as imperfect knowledge about the prerequisites necessary. These imperfections in the knowledge of the prerequisites include:

- uncertainty in TOF introduced by inaccurate initial position approximation

- imprecision in code phase measurement due to limited sampling rate and RF frontend limitations (especially filter characteristic)

- influence of the radio channel

Influence of Initial Position Approximation on Time Information

Remembering chapter 5 the positioning operation in the proposed architecture requires an initial position approximation based on an attached cellular network connection. Depending on the type of cellular network the cells cover radii between hundreds of meters in case of WLAN up to approximately 20 km for GSM cells. As already stated the proposed algorithm equates the position of the cell's base station with the approximate user position. Hence the worst case assumptions for the distance between real (Points A, B) and estimated position (C) as illustrated in Figure 6.6 are acceptable.

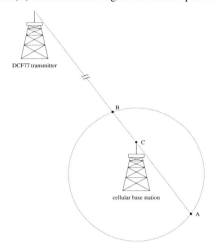

Figure 6.6: Worst-case assumption for inital position approximation error

For easy integration into the existing system simulation positions are switched vice versa, assuming point C to be the real user position and implementing worst-case considerations by shifting the initial position approximation to points A and B. Calculating points A and B exemplary assuming the real user position to be at 49.573385 ° N, 11.029109 ° E leads to the positions for A,B and C as given in table 6.1 and illustrated in Fig. 6.7.

Point	Latitude	Longitude
A	49.543352838803020 ° N	11.159365315038500 ° E
B	49.603271465905880 ° N	10.898689424400180 ° E
User Position (C)	49.573385 ° N	11.0291098333333 ° E

Table 6.1: Exemplary worst-case error assumption positions

The possible error becomes manifest in different phases of the maximum of the cross correlation function between the received signal and its local generated replica when simu-

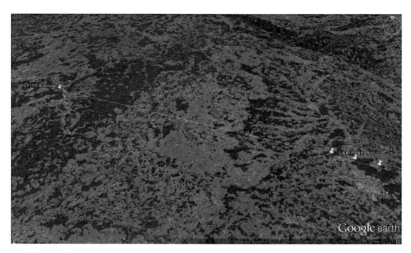

Figure 6.7: Exemplary worst-case calculation for possible inital position approximation

lated for worst-case position approximation assumptions. The results are given in Figures 6.8.

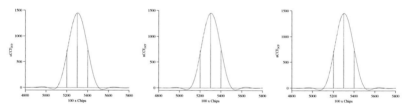

Figure 6.8: Output of cross correlation function for worst case approximation of initial position (solutions for points A (red), B(blue) and C(green) from Figure 6.6

It becomes clear that the differences in code phase cannot be resolved with the architecture at hand for worst case initial position assumptions. The resolution which must be achieved can be approximated by equations (6.12), leaving the error in TOF due to the distance between approximate and real position as remainder in time synchronization. This synchronization error of the internal clock versus UTC is summarized for positions A,B and C in table 6.2.

$$
\begin{aligned}
t_{A-B} &= \frac{d_{A-B}}{c} = \frac{20\,\mathrm{km}}{299792.458\,\frac{\mathrm{km}}{\mathrm{s}}} = 66.712819\,\mathrm{\mu s} \\
t_{\Delta} &= \frac{t_{epoch\_DCF77}\,\Delta_{channel\_spacing}}{n_{chips}\,n_{interpolation\_os}} = \frac{1\,\mathrm{s}\,0.5}{646\,100} = 7.74\,\mathrm{\mu s}
\end{aligned}
\tag{6.12}
$$

	Calculated PRN Shift [chips]	Real PRN Shift	Error [chips]	Error [µs]
Point A	52.741139	53.918372	-1.177233	-356.30
Point B	52.719603	53.918372	-1.198769	-322.94
Point C	52.698073	53.918372	-1.2203	-305.08

Table 6.2: Worst-case time errors compared to approximate inital position

Influence of Distortion in Radio Channel on Time Information

The simulation results presented in Fig. 6.8 only cover issues because of inaccuracy of initial position approximation. For the simulation of the influence of the radio channel the modified model introduced in section 6.1.1 comes into play. Figure 6.9 shows simulation results of the baseband correlation engine neglecting any noise or other distortion in the radio channel at an artificial phase shift of -150 chips.

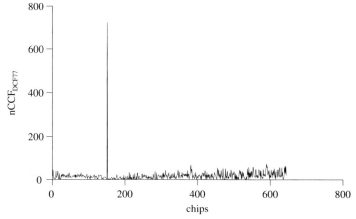

Figure 6.9: Correlation output, applying DCF77 RF-Frontend simulation with no distortion effects

Simulating reception conditions which are rarely present to the receiver, still good correlation results of the local PRN and the received signal can be achieved, as illustrated in Fig. 6.10.

6.2.2   GNSS Simulation Results

This section presents the results from the system simulation trying to highlight the influence of different parameters on the positioning solution. Although there are plenty

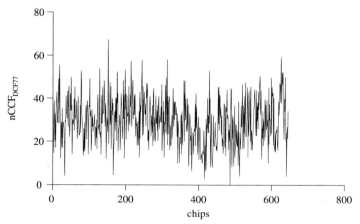

Figure 6.10: Correlation output, applying DCF77 RF-Frontend simulation with .1% jitter on LO input of mixer, -25 dB SNR and interference signal with 3 dB higher signal power than DCF77 SIS

parameters influencing the accuracy of the positioning solution three are identified for the architecture at hand to be most important, which will be investigated:

- initial position approximation

- DCF77 signal strength

- NAVSTAR GPS signal attenuation

Main focus of the simulation setup is to prove functionality of the proposed architecture as well as identification of conditions when the positioning algorithm collapses. Initially, results of a simulation setup demonstrating an undisturbed scenario, to be used as reference, are presented. Following the introductory position approximation is again varied to worst-case assumptions, similar to the DCF77 standalone simulation. It will be shown that there is negligible influence on the position solution. Next different types and levels of disturbances, in particular noise, for the DCF77 and NAVSTAR GPS signals are added to the simulation setup and the influence on the simulation results are investigated. These scenarios reflect different non-ideal reception conditions, e.g. limited view to the sky.

Undisturbed Scenario

According to the postulated architecture, simulation of the receiver operation should deliver a maximum peak in the output of the cross correlation function for each SV. These serve as basis for the determination of the pseudoranges to the SVs and consequently the positioning solution. Assuming a clear sky view, hence good reception conditions for DCF77 and NAVSTAR GPS signal reception and a good accordance between both, real- and approximated user position, it is inevitable to find the desired maximum in the search space. The maximum of the output of the cross correlation function should not only be present but additionally include a high peak-to-average ratio.

The boundary conditions for the simulation of an almost undisturbed scenario, reflecting good reception conditions for GNSS as well as DCF77 signals, are given by the parameters in table 6.3.

real user position	49.573385 °E	11.0291097 °N	292 m
approximate user position	49.573385 °E	11.0291098 °N	292 m
date and time		April 8th 2014	10:32:45.0830
SNR_{DCF77}			10 dB
DCF77 coherent integration time			1 s
$\frac{C}{N_0}$			-21 dB
GNSS coherent integration time			4 ms

Table 6.3: Prerequisites of simulation in ideal reception environments

Exemplary output of the cross correlation bank for NAVSTAR GPS SVs in view, implementing the listed boundary conditions, are given in Figure 6.11. It has to be mentioned that the output vectors of the cross correlation bank are extended to the nominal number of channels needed in non-time-assisted architectures and shifted to the corrected by the PRN phase approximation for clarity reasons. The raw output vectors all occupy the same phase shift position with their maximum. This behavior is inherent to the architecture due to the application of the precalculated approximation of the phase shift to the local PRN replica prior to the cross correlation operation. The presented results shown in Fig. 6.11 are already evaluated by the sinc-interpolator.

For better illustration, raw CCF results and the corresponding output of the sinc-interpolator for SV 2 are given in Figure 6.12 together with a focused view on the range around the detected maximum in 6.13. Special attention must be paid to the fact, that the peak of the result is located in the middle of the correlation bank. An additional shift of half the number of channels in the cross correlation bank is added to the local replica of the PRN sequence. This gives two advantages over a straight forward approach without additional shift. Depending on the initial position approximation both, overestimation and underestimation of the current code phase is possible. The straight forward approach can only cover underestimation of the code phase, while an overestimation would shift the peak of the cross correlation function outside of the search window covered by the correlation bank. Second, interpolation techniques for elimination of the error introduced by the finite code phase distance between adjacent correlation channels, hence the determination of the real cross correlation maximum, using recursion can deliver better results if already tuned when running over the peak results.

The resulting simulated code phases for the NAVSTAR GPS satellites in view and their nominal counterparts are given in table 6.4.

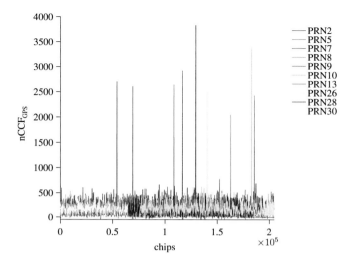

Figure 6.11: Output of CCF for SVs in view at boundary conditions given in table 6.3 corrected by approximated phase shift of local PRN replica

SV	distance[km]	code phase (ideal)	code phase (simulated receiver)
2	24.231,77	270,79525	270,78000
5	20.855,77	542,77325	542,73000
7	21.220,84	346,27400	346,26500
8	20.495,55	814,67200	814,51000
9	19.925,52	584,40025	584,27500
10	22.021,25	915,43075	915,25500
13	23.895,10	928,06025	928,00000
26	21.695,04	164,29500	164,28000
28	23.032,78	647,35850	647,23500
30	20.261,99	701,88525	701,73500

Table 6.4: Simulated code phases for SVs in view assuming conditions according table 6.3

The results of the positioning solution for five subsequent runs are averaged and illustrated in Figure 6.14 and listed in table 6.5, respectively. Additionally the nominal code phases used for simulation of the SISs are listed, too.

As expected both, calculated and real positioning solutions, differ due to noise and the limited bandwidth of the receiver as well as the limited sampling rate. The discrepancy between real position and the mean position calculated from the raw signals at the receiver

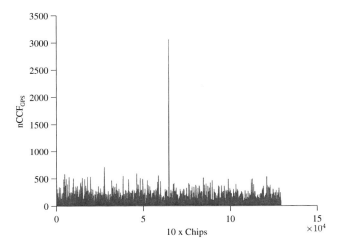

Figure 6.12: Output of cross correlation bank (coherent integration time: 4 ms) for SV 2, 10 dB SNR for DCF77 signal and $\frac{C}{N_0}$ = -21 db for GPS (red: output of correlation bank, blue: x100 sinc interpolation)

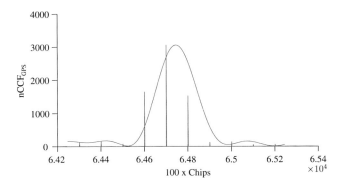

Figure 6.13: Focus on peak correlation output as given in Fig. 6.12 (red: output of correlation bank, blue: x100 sinc interpolation)

are originates from the numeric algorithm implemented to solve the system of equations describing the positioning solution. Nevertheless both solutions show good consistency, hence assuming a clear sky the presented approach delivers good positioning results.

Figure 6.14: Preconditions and results (mean of 5 iterations) of positioning solution in an undisturbed scenario (corners of white line: individual results)

approximate position	mean calculated position	real position	mean error [m]
49,573385° E 11,029109° N	49,573289° E 11,029003° N	49,573385° E 11,029109° N	20,28

Table 6.5: approximate versus calculated versus real user position

Worst-case Faulty Initial Position Approximation Scenario

As stated earlier the approximate position at the beginning of the algorithm is set to the position of the base station the cellular network connection is logged onto. It has already been proven that time information error introduced by faulty initial position approximation to the DCF77 receiver does not have significant effect on the time information gathered. Nevertheless it is inevitable to evidence that the combined error of degraded time information and consequently error in the precalculation of the NAVSTAR GPS code phases do not exceed the search window provided by the correlation bank in the receiver. Therefore two worst-case scenarios are imaginable as already illustrated in Figure 6.6 in section 6.2.1.

The maximum error in time synchronization is always present at the intersections A and B of the straight line running through the DCF transmitter and the position of the cell's base station with an imagined circle having the radius of the maximum coverage of the cellular network cell. For common applications this diameter can be approximated with 20 km, assuming a large GSM cell.

The following assumptions for the environment during simulation were applied:

	Point A	Point B
real user position	49,573385° E	49,573385° E
	11,0291097° N	11,029109° N
approximate user position	49,603271° E	49,543352° E
	10,8986894° N	11,159364° N
date and time	April 8th 2014, 10:32:45.0830	
SNR_{DCF77}	10 dB	
DCF77 coherent integration time	1 s	
$\frac{C}{N_0}$	-21 dB	
GNSS coherent integration time	4 ms	

Table 6.6: Prerequisites of simulation for Points A and B as given in Figure 6.6

The resulting simulated code phases for the NAVSTAR GPS satellites SV 2 and SV 5 in view and their real counterparts are given in table 6.7.

SV 2	real code phase	simulated code phase	estimated code phase
Point A	270,795250	270,760000	290
Point B	270,795250	270,770000	251
SV 5	real code phase	simulated code phase	estimated code phase
Point A	542,773250	542,720000	563
Point B	542,773250	542,750000	522

Table 6.7: Real versus simulated versus estimated code phases for SV 2 and 5

As expected, the estimated code phases exhibit considerably more code phase distance from the simulated results. But the distance in code phase do not exceed the search space covered by the architecture proposed. The code phase distance introduced, is the super-position of several individual error sources whose sum must not exceed the code phase search window. The imprecise distance approximation to the DCF77 transmitter results in an inaccurate TOF compensation for the DCF77 signal. This error together with the imprecise approximation of the current position lead to a degraded forecast of the expected GNSS code phases. This corresponds to the considerations for software realization in section 5.5. Certainly the calculated position error does not significantly differ from the undisturbed scenario. Hence even greater distances to the cell's base station can be covered as long as the cellular network's range is not exceeded or the approximated position and assistance data is gathered otherwise. The limited accuracy of the calculated code phases in table 6.7, to two decimals is due to the oversampling factor of $k_{os\_sinc} = 100$ for the sinc-interpolation technique. Obviously the algorithm is not able to determine the interpolated maximum of the CCF more precise than the resolution restricted to $\frac{1}{k_{bs\_sinc}}$.

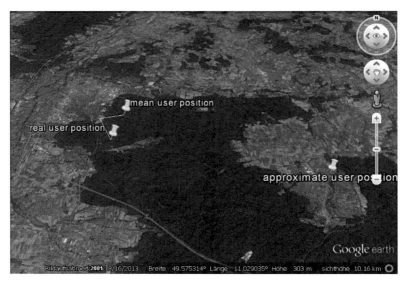

Figure 6.15: Preconditions and results (mean of 5 iterations) of positioning solution in an undisturbed scenario (corners of white line: individual results)

Point A approximate position	mean calculated position	real position	mean error [m]
49,573385° E 11,029109° N	49,573289° E 11,029003° N	49,573385° E 11,029109° N	20,28
Point B approximate position	mean calculated position	real position	mean error [m]
49,603271° E 10,898689° N	49,573279° E 11,029069° N	49,573385° E 11,029109° N	19,93

Table 6.8: approximate versus calculated versus real user position for maximum faulty position approximation

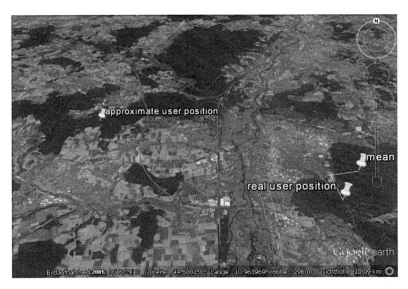

Figure 6.16: Preconditions and results (mean of 5 iterations) of positioning solution in an undisturbed scenario (corners of white line: individual results)

Influence of DCF77 Signal Quality on Positioning Solution

Obviously the proposed architecture looses its superiority in reception environments where the assistance by time information is not available, hence the current code phase of the DCF77 signal cannot be determined. However results of the cross correlation function with a low peak-to-average ratio degrade the time information gathered, too. Again this has direct impact on the forecast of the GNSS code phase because of inaccurate interpolation results. Fortunately the error in time information cannot exceed half the code distance between two adjacent cross correlation channels. Consequently the maximum error in time synchronization, remembering the considerations in 5.5 and the DCF77 PRN parameters (length of PRNsequence n_{epoch} and time length of epoch t_{epoch}), should not exceed approximately $387{,}2\,\mu s$ according to equation 6.13.

$$t_{err,max} = \frac{1}{2}d_{channels}\frac{t_{epoch}}{n_{epoch}} = \frac{1}{2}0.5\frac{793\,\text{ms}}{512} \approx 387{,}2\,\mu s \qquad (6.13)$$

Setting up an simulation environment, implementing the Signal-to-Noise Ratio (SNR) value for the DCF77 signal where the time information cannot be gathered anymore does not deliver new expertise, since it marks the border when time synchronization cannot be guaranteed due to a missing peak in the cross correlation spectrum as well as a collapse of the whole algorithm. As given in Figure 6.17 the degradation of the peak value of the cross correlation function depending on the DCF77 signal's SNR can be observed.

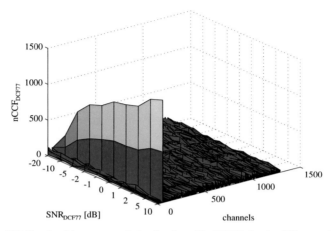

Figure 6.17: Results of the cross correlation function of the DCF77 signal at different noise levels

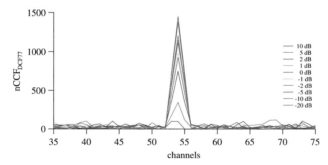

Figure 6.18: Maxima of the results of the cross correlation function of the DCF77 signal at different noise levels

The calculated time information for several noise levels of the DCF77 signal, together with the calculated code phases for SV 2, are given in table 6.9

Looking at the results given in table 6.9 and Figure 6.19 it becomes clear, that as long as the SNR of the DCF77 signal is high enough to allow the determination of the maximum of the cross correlation function the influence on the positioning solution can be neglected. This approach is legal since the maximum error introduced by the inaccuracy in time

DCF77 noise level [dB]	DCF77 code phase [chips]	calculated time April 8th, 2014	calculated code phase GPS SV 2 [chips]
10	52,70	10:32:45,082661	270,785
5	52,70	10:32:45,082661	270,780
2	52,71	10:32:45,082677	270,790
1	52,74	10:32:45,082723	270,755
0	52,72	10:32:45,082692	270,770
-1	52,72	10:32:45,082692	270,810
-2	52,73	10:32:45,082708	270,780
-5	52,69	10:32:45,082646	270,760
-10	52,64	10:32:45,082568	270,780
-20	434,89	10:32:45,674608	270,760

Table 6.9: Time synchronization and calculated code phases for DCF77 signal at several noise levels compared to calculated code phase for NAVSTAR GPS SV 2

Figure 6.19: Results of the cross correlation function of the DCF77 signal at different noise levels

synchronization to the estimation of the GNSS code phases is smaller than the search window covered by the cross correlation bank as defined in section 5.

Influence of NAVSTAR GPS Signal Quality on Positioning Solution

Obviously the most important factor for calculating the positioning solution is the avail-ability of the GNSS signals at the receiver and their signal quality respectively. All oper-ations prior to the cross correlation of the NAVSTAR GPS signals are obsolete in case of extreme bad reception environments, hence very low $\frac{C}{N_0}$.

All calculations, synchronizing time information and precalculation of approximated GNSS code phases target the efficiency of the receiver, hence reduce the amount of operations necessary to find the maximum of the cross correlation function. Consequently the pro-posed architecture deals with the same boundary conditions as every common GNSS receiver architecture. Assuming all boundary conditions, as described in the preceding subsections, are met and a consequently correct approximations of the NAVSTAR GPS code phases are possible, the output of the cross correlation function deteriorates with degrading $\frac{C}{N_0}$ as illustrated exemplary for GPS SV 2 in Figure 6.20.

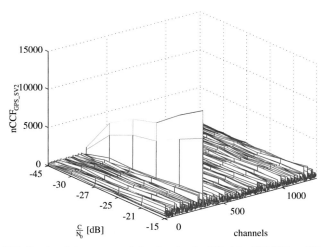

Figure 6.20: Results of the cross correlation function for SIS of the GPS SV2 at different noise levels (Bandwidth= 4 MHz)

Zooming the peak results of the CCF from Fig. 6.20, the degradation of the code phase solution to the point when no peak in the CCF solution is identifiable anymore gets more obvious in Fig. 6.21.

The direct impact on the positioning solution can be illustrated in Fig. 6.22.

While peak results are identifiable in the output of the CCF, results of the positioning solution have small deviation from each other and are located near the real position of the user. As soon as the the peaks in the CCF results disappear the positioning algorithm no longer gives significant results.

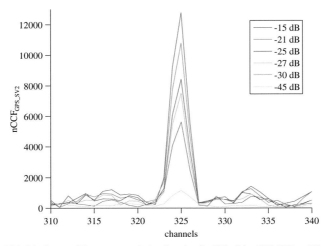

Figure 6.21: Maximum of the cross correlation function for SIS of the GPS SV2 at different noise levels (Bandwidth= 4 MHz)

The corresponding numbers, illustrated by Fig. 6.22 and hence the development of the position solution precision for several $\frac{C}{N_0}$ values can be observed in table 6.10.

$\frac{C}{N_0}$ [dB]	real user position	calculated user position	position error [m]
-15	49,573385° N	49,5732789° N	19,417
	11,029109° E	11,0292555° E	
-21	49,573385° N	49,5733339° N	19,230
	11,029109° E	11,0292692° E	
-25	49,573385° N	49,5732777° N	19,148
	11,029109° E	11,0292321° E	
-27	49,573385° N	49,5733340° N	19,125
	11,029109° E	11,0292375° E	
-30	49,573385° N	49,5733571° N	23,193
	11,029109° E	11,0292611° E	
-45	49,573385° N	49,5733996° N	44,415
	11,029109° E	11,0294036° E	

Table 6.10: Positioning solution and position error plotted versus noise level and real user position

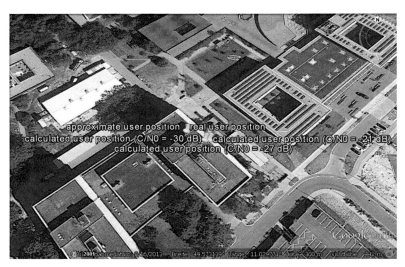

Figure 6.22: Calculated user position compared to real and approximate user position prior for different noise levels for GPS signals, including faulty reception conditions

The rest is silence.

_____
William Shakespeare

CHAPTER 7

Summary and Outlook

For the first time an assisted architecture for GNSS receivers with focus on implementation in mobile devices has been described and characterized, incorporating the information from a time code receiver. The preceding chapters deeply investigated the basics of satellite navigation and its sensitivity to environmental parameters. The drawbacks of common GNSS receiver architectures either assisted or not where highlighted. These constitute starting points which lead to the herein proposed essential architectural changes to overcome the mentioned challenges. After discussion of a possible realization of the new architecture the results of intensive system simulations are presented proving functionality as well as superiority over established GNSS receiver solutions.

The new developed architecture conserves the advantages of common assisted navigation receivers, namely short TTFF and significant reduction in power consumption, compared to the ATR architecture. Additionally it eliminates the main disadvantage of assisted receiver architectures, a massive usage of computational power in the digital baseband domain.

Common to assisted architectures for navigation receivers, the proposed implementation of an GNSS receiver exploits the synergies of integrating a cellular network connection in parallel with the positioning capabilities. Additionally the sensor data fusion is extended by time information via a LW time code signal. The navigation message can, in contrast to, e.g., the ATR, be made available via the cellular data connection. Starting with a rough estimation of time and position based again on the cellular connection, the navigation message gives instant knowledge about satellites in-view as well as a hint about the Doppler drift of the incoming satellite signals. A massive parallel correlation architecture is applied to the incoming signal performing the cross correlation function with a local generated replica. The amount of processing power needed to calculate is dependent different parameters like sampling rate, channel distance or length of the PRN sequence. Assuming no a priori knowledge about the expected code phase this correlation bank can implement a significant amount of hardware. The herein demonstrated architecture avoids this problem by the integration of precise time information into the positioning solution. Precise knowledge about reception time of the GNSS signal allows a calculation of the satellite constellation. A position approximation, whose deviation from the real user position does not exceed on code epoch allows precalculation of the prospective codephase between received signal and local generated replica. This gives the opportunity to reduce the amount of cross correlation channels for determining the exact code phase of the appropriate satellite signals, depending on the quality of the time information.

Responsible for generating the information is a DCF77 time code receiver. Fortunately as described in chapters 4 and 5 the DCF77 SIS has a similar information structure as the GNSS signals. Both transfer information via PRN sequences. Reusing the correlation bank for determination of the time information and keeping in mind to that the DCF77 code cycle has only half the length of NAVSTAR GPS PRN sequences and even $\frac{1}{8}$ of those of Galileo, a cut in effort nearly reaches a factor of two for NAVSTAR GPS and eight for Galileo, respectively. Nevertheless positioning accuracy does not and TTFF does only slightly degrade compared to other assisted GNSS receiver architectures. Even in the absence of any cellular network connection the proposed architecture allows superior positioning times as far as an approximate position is known.

Still there is much potential for further development of the herein proposed architecture. First of all the amount of processing channels to be provided is constrained to its lower bound by the amount of channels needed to scan the whole DCF77 code phase epoch for code phase equality. It is imaginable to apply the idea of precalculation a second time. Implementing networking protocols like NTP or Precision-Time-Protocol (PTP) to the handset could deliver information about approximate current time. This information can help to precalculate the actual code phase of the DCF77 signal and further cut down the amount of parallel processing channels in the cross correlation bank. Second the full integration of the usage of Galileo signals has to be implemented. With the restored full constellation of the Russian GLONASS system and the integration of its signals into the receiver architecture would allow to raise the number of satellites available further. Especially in bad reception conditions, e.g. in urban canyons, the possibility of line-of-sight to 4 satellites would be enhanced considerably. Still the architecture would benefit from its reduced hardware usage, since integration of GLONASS mainly concerns the availability of the local generated replica. From that even more Global Navigation Satellite Systems can easily be integrated.

Additionally, up to now the presented architecture is limited to a channel stepping of $\frac{1}{2}$ chip when calculating the CCF for GNSS signals. But however there is no limitation to adapt the channel spacing if there is a figure of merit available describing the accuracy of the time information gathered via the DCF77 signal.

Bibliography

[1] [Online]. Available: http://www.gps.gov/multimedia/images/

[2] Navipedia, "GNSS Signal — Navipedia," 2011, online; accessed 3-November-2013. [Online]. Available: http://www.navipedia.net/index.php?title=GNSS_signal&oldid=8875

[3] [Online]. Available: www.esa.int

[4] [Online]. Available: http://www.navipedia.net/index.php/CBCS_Modulation

[5] P. Hetzel, "Zeitübertragung auf Langwelle durch amplitudenmodulierte Zeitsignale und pseudozufällige Umtastung der Trägerphase," Ph.D. dissertation, Institut für Zeitmeßtechnik, Fein- und Mikrotechnik der Universität Stuttgart, 1987.

[6] [Online]. Available: http://www.conrad.de/ce/de/product/641138/Conrad-DCF-Empfaengerplatine

[7] "Global Positioning System Standard Positioning Service Signal Specification," vol. 2nd Ed., June 1995.

[8] D. Piester, P. Hetzel, and A. Bauch, "Zeit- und Normalfrequenzverbreitung mit DCF77," Physikalisch Technische Bundesanstalt, Tech. Rep. 114 Heft 4, 2004.

[9] The White House, Office of the Press Secretary, "Statement by the President Regarding the United States Desision to Stop Degrading Global Positioning System Accuracy," May 1st, 2000.

[10] D. Akopian, "A fast satellite acquisition method," *Institute of Navigation Conference ION GPS 2001*, pp. 2871 – 2881, September 2001.

[11] N. F. Krasner, "Method and Apparatus for Determining Time for GPS Receivers," U.S. Patent 5 945 944, 1999.

[12] G. Heinrichs, N. Lemke, A. Schmidt, A. Neubauer, R. Kronberger, G. Rohmer, F. Förster, T. Pany, J. A. Avila-Rodriguez, B. Eisfeller, H. Ehm, and R. Weigel, "HI-GAPS- A Large-Scale Integrated Combined Galileo/GPS Chipset for the Consumer Market," in *Proceedings of ENC-GNSS*, 2004.

[13] E. D. Kaplan and C. J. Hegarty, *Understanding GPS: Principles and Applications*, 2nd ed.   Norwood: Artech House, Inc., 2006.

[14] *NAVSTAR GPS Space Segment/Navigation User Interfaces, IS-GPS-200D*, ARNIC Engineering Services, March 7th, 2006.

[15] J. L. Leva, "An Alternative Closed-Form Solution to the GPS Pseudo-Range Equations," *IEEE Transactions on Aerospace and Electronic Systems*, vol. 32, no. 4, pp. 1430 –1439, October 1996.

[16] S. Bancroft, "An Algebraic Solution of the GPS Equations," *IEEE Transactions on Aerospace and Electronic Systems*, vol. AES-21, no. 1, pp. 56 –59, January 1985.

[17] B. Hofmann-Wellenhof, H. Lichtenegger, and J. Collins, *Global Positioning System: Theory and Practice*.   Springer-Verlag, 1997.

[18] K. Davies, *Ionospheric Radio*, ser. IEE Electromagnetic Waves Series.   The Institution of Engineering and Technology, 1990, vol. 31.

[19] J. A. Klobuchar, "Ionospheric Time-Delay Algorithm for Single-Frequency GPS Users," *IEEE Transcations on Aerospace and Electronic Systems*, vol. AES-23, no. 3, pp. 325 –331, May 1987.

[20] ——, "Ionospheric Corrections for Timing Applications," *Proceedings of the Twentieth Annual Precise Time and Time Interval (PTTI) Applications and Planning Meeting*, 1988.

[21] *Minimum Operaional Performance Standards For Global Positioning System/Wide Area Augmentation System Airborne Equipment*, Radio Technical Commission for Aeronautics Std. DO-229C, November 2001. [Online]. Available: http://www.rtca. org/doclist.asp

[22] J. Guo and R. B. Langley, "A New Tropospheric Propagation Delay Mapping Function for Elev. Angles Down to 2 deg," in *Proceedings of ION GPS/GNSS 2003, 16th International Technical Meeting of the Satellite Division of The Institute of Navigation*.   The Institute of Navigation, September 2003, pp. 386–396.

[23] [Online]. Available: http://www.navcen.uscg.gov/?pageName=GPSmain

[24] R. Gold, "Optimal Binary Sequences for Spread Spectrum Multiplexing (Corresp.)," *IEEE Transactions on Information Theory*, vol. 13, no. 4, pp. 619 – 621, October 1967.

[25] S. W. Golomb, *Shift Register Sequences*.   Aegean Park Pr, 1981.

[26] *Agreement on the Promotion, Provision and Use of Galileo and GPS Satellite-Based Navigation Systems and Related Applications*, EU and USA, June 28th, 2004.

[27] *United States and the European Union Announce Final Design for GPS-Galileo Common Civil Signal*, EU press release IP/07/1180, June 27th, 2007.

[28] J. G. Walker, "Satellite Constellations," *Journal of the British Interplanetary Society*, vol. 37, pp. 559–572, December 1984.

[29] European Commision, "Galileo Mission High Level Definition," Tech. Rep., September 2002. [Online]. Available: http://ec.europa.eu/dgs/energy_transport/galileo/doc/galileo_hld_v3_23_09_02.pdf

[30] *European GNSS (Galileo) Open Service Signal In Space Interface Control Document*, European Union, 2010.

[31] J. W. Betz, "Binary Offset Carrier Modulations For Radionavigation," *Navigation, Journal of The Institute of Navigation*, vol. 48, pp. 227–246, 2001.

[32] H. Ehm, "Single Shot Radio Receiver Architectures for Mobile Stations," Ph.D. dissertation, Lehrstuhl für Technische Elektronik der Universität Erlangen-Nürnberg, 2008.

[33] B. Parkinson and J. Spilker, *Global Positioning System: Theory and Applications*, ser. Global Positioning System: Theory and Applications. American Institute of Aeronautics and Astronautics, 1996, no. Bd. 1. [Online]. Available: http://books.google.de/books?id=lvI1a5J_4ewC

[34] *Location Services (LCS) Technical Specifications 3GPP TS 04.30, 04.31, 24.030, 44.031, 44.035, 44.071*, 3rd Generation Partnership Project, 2003.

[35] G. Heinrichs, N. Lemke, A. Schmidt, A. Neubauer, R. Kronberger, G. Rohmer, F. Förster, J. A. Avila-Rodriguez, T. Pany, B. Eisfeller, H. Ehm, and R. Weigel, "Galileo/GPS Receiver Architecture for High Sensitivity Acquisition," in *Proceedings of International Symposium on Signals, Systems and Electronics*, 2004.

[36] D. Benson, "GPS L1 C/A Signal Acquisition Analysis," RTCA, Tech. Rep. RTCA SC-159 WG-6, September 2006.

[37] A. J. V. Dierendonck, P. Fenton, and T. Ford, "Theory and Performance of Narrow Correlator Spacing in a GPS Receiver," in *Proceedings of the 1992 National Technical Meeting of The Institute of Navigation*, no. 3. The Institute of Navigation, January 1992, pp. 115 – 124.

[38] R. Zheng, M. Chen, X. Ba, and J. Chen, "A Novel Fine Code Phase Determination Approach for a Bandwidth Limited Snapshot GPS Receiver," in *IEEE/ION Position Location and Navigation Symposium (PLANS) 2010*, May 2010, pp. 796 –805.

[39] N. Sirola and J. Syrjärinne, "GPS Position Can Be Computed Without Navigation Data," in *Proceedings of the ION GPS 2002*, 2002, pp. 2741–2744.

[40] T. Pany, E. Göhler, M. Irsigler, and J. Winkel, "On the State-of-the-Art of Real-Time GNSS Signal Acquisition; A Comparison of Time and Frequency Domain Methods," in *International Conference on Indoor Positioning and Indoor Navigation (IPIN) 2010*, 2010, pp. 1–8.

[41] D. Manandhar, Y. Suh, and R. Shibasaki, "GPS Signal Acquisition and Tracking-An Approach Towards Development of Software-based GPS Receiver," *Tokyo, The Institute of Electronics, Information and Communication Engineers, Technical Report*, 2004.

[42] ITU, *RECOMMENDATION ITU-R TF.460-6: Standard-Frequency and Time-Signal Emissions*, International Telecommunications Union, 2002.

[43] Bundesamt für Metrologie METAS, "Zeitzeichensender HBG 75 KHz," Eidgenössisches Justiz- und Polizeidepartement EJPD Bundesamt für Metrologie METAS Labor für Zeit und Frequenz, Tech. Rep., January 2007.

[44] National Physical Laboratory, "NPL Time & Frequency Services - MSF 60 kHz Time and Date Code," National Physical Laboratory Industry & Innovation Division Time and Frequency Services, Tech. Rep., July 2007.

[45] M. Lombardi, "NIST Time and Frequency Services," *NIST Special Publication*, vol. 432, 2002.

[46] K. Betke, "Standard Frequency And Time Signal Stations On Longwave And Shortwave," Institut für Technische und Angewandte Physik, Tech. Rep., August 2006.

[47] [Online]. Available: http://www.heret.de/funkuhr/reichw.htm

[48] A. Bauch, P. Hetzel, and D. Piester, "Zeit- und Frequenzverbreitung mit DCF77: 1959 – 2009 und darüber hinaus," Physikalisch Technische Bundesanstalt, Tech. Rep. 119 Heft 3, 2009.

[49] C. A. Balanis, *Antenna Theory: Analysis and Design*, 3rd ed.   Wiley-Interscience, 2005.

[50] W. Hilberg, *Funkuhrtechnik*.   München: Oldenbourg Verlag GmbH, 1988.

[51] *Technical Information Operating Instructions PZF 600*, Meinberg Solutions for Time and Frequency Synchronization.

[52] C. Shannon, "Communication in the Presence of Noise," *Proceedings of the IEEE*, vol. 72, no. 9, pp. 1192–1201, 1984.

[53] *RapidIO MegaCore Function User Guide*, 13th ed., Altera Inc., May 2013.

[54] A. Schmidt, A. Neubauer, H. Ehm, R. Weigel, N. Lemke, G. Heinrichs, J. Winkel, J. A. Avila-Rodriguez, R. Kaniuth, T. Pany, B. Eisfeller, G. Rohmer, B. Niemann, and M. Overbeck, "Enabling Location Based Servies with a Combined Galileo/GPS Receiver Architecture," in *Proceedings of ION GNSS*, 2004.

[55] A. Schmid and H. Ehm, "Hoch-Integrierter Galileo /GPS Empfänger Chipsatz Phase 1 Schlussbericht," July 2006.

[56] S. Furber, *ARM System-on-Chip Architecture.* Addison-Wesley, 2000. [Online]. Available: http://books.google.de/books?id=wHERBxn0tH0C

[57] R. Islam, A. Sabbavarapu, and R. Patel, "Power Reduction Schemes in Next Generation Intel ATOM Processor Based SoC for Handheld Applications," in *VLSI Circuits (VLSIC), 2010 IEEE Symposium on*, 2010, pp. 173–174.

[58] [Online]. Available: http://www.android.com/

[59] M. Sauter, *From GSM to LTE: An Introduction to Mobile Networks and Mobile Broadband*, ser. Wiley Online Library: Books. Wiley, 2010.

[60] [Online]. Available: http://www.wolframalpha.com/input/?i=how+fast+do+gps+satellites+travel

[61] D. Mills, J. Martin, J. Burbank, and W. Kasch, "Network Time Protocol Version 4: Protocol and Algorithms Specification," RFC 5905 (Proposed Standard), Internet Engineering Task Force, Jun. 2010. [Online]. Available: http://www.ietf.org/rfc/rfc5905.txt

[62] *Information Technology - Open Systems Interconnection - Basic Reference Model: The Basic Model ITU-Z RECOMMENDATION X.200 ISO/IEC 7498-1:1994*, TELECOMMUNICATION STANDARDIZATION SECTOR OF ITU Std.

[63] J. Whittaker, *Interpolatory Function Theory*, ser. Cambridge Tracts in Mathematics and Mathematical Physics. The University Press Cambridge, 1935, no. 33.

[64] W. Gurtner, G. Mader, and D. MacArthur, "A Common Exchange Format for GPS Data," in *Proceedings of the Fifth International Geodetic Symposium on Satellite Systems*, Las Cruces., 1989, p. 917ff.

[65] H. D. Black and A. Eisner, "Correcting Satellite Doppler Data for Tropospheric Effects," *Journal of Geophysical Research: Atmospheres*, vol. 89, no. D2, pp. 2616–2626, 1984.

[66] K. Borre, D. M. Akos, N. Bertelsen, P. Rinder, and S. H. Jensen, *A Software-Defined GPS And Galileo Receiver - A Single-Frequency Approach*, 2007th ed. Berlin, Heidelberg: Springer, 2007.

[67] B. Tolman, R. B. Harris, T. Gaussiran, D. Munton, J. Little, R. Mach, S. Nelsen, and B. Renfro, "The GPS Toolkit: Open Source GPS Software," in *Proceedings of the 16th International Technical Meeting of the Satellite Division of the Institute of Navigation*, Long Beach, California, September 2004.

[68] P. Hetzel, "Der Langwellensender DCF77 auf 77,5 kHz: 40 Jahre Zeitsignale und Normalfrequenz, 25 Jahre kodierte Zeitinformation," *PTB Mitteilungen*, vol. 109, pp. 11–18, 1999.

[69] *Frequenznutzungsplan gemäß §54 TKG über die Aufteilung des Frequenzbereichs von 9 kHz bis 275 GHz auf die Frequenznutzungen sowie über die Festlegungen fü r diese Frequenznutzungen*, Bundesnetzagentur Std., 2011.

[70] H. Friis, "A Note on a Simple Transmission Formula," *Proceedings of the IRE*, vol. 34, no. 5, pp. 254–256, 1946.

[71] "Technology - Antenna Design," C-MAX Time Solutions GmbH, Tech. Rep. [Online]. Available: http://www.c-max-time.com/tech/antenna.php

[72] [Online]. Available: ftp://cddis.gsfc.nasa.gov/pub/gps/data/hourly

[73] C. Gauß, *Theorematis Arithmetici Demonstratio Nova*, ser. Werke, 1863. [Online]. Available: http://books.google.de/books?id=A0nwNAEACAAJ

[74] J.-M. Zogg, *GPS und GNSS: Grundlagen der Ortung und Navigation mit Satelliten.* online, 2011.

[75] H. Friis, "Noise Figures of Radio Receivers," *Proceedings of the IRE*, vol. 32, no. 7, pp. 419–422, 1944.

Own Publications

C. Kandziora and R. Weigel, „Time-Code Assisted Low-Power GNSS Single-Shot Receiver for Mobile Devices", International Frequency Control and the European Frequency and Time Forum (IFCS), 2011 Joint Conference of the IEEE, 2011.

C. Kandziora and R. Weigel, „Concept for a Time-Code Signal Assisted Single-Shot Receiver", Radio and Wireless Symposium (RWS), IEEE, 2011.

C. Kandziora and R. Weigel, „Time Signal Assisted GPS Receiver for Mobile Devices", Proceedings of the 23rd International Technical Meeting of The Satellite Division of the Institute of Navigation (ION GNSS 2010), pp. 29772980, September 2010

M. Jank, C. Kandziora, L. Frey and H. Ryssel, „Well Design In A Bulk CMOS Technology With Low Mask Count", AIP Conference Proceedings, vol. 866, no. 1, pp. 121-124, 2006